Brown and Sharpe Manufacturing Co.

A Treatise on the Construction and Use of Milling Machines

Made by Brown & Sharpe mfg. co.

Brown and Sharpe Manufacturing Co.

A Treatise on the Construction and Use of Milling Machines
Made by Brown & Sharpe mfg. co.

ISBN/EAN: 9783337163723

Printed in Europe, USA, Canada, Australia, Japan

Cover: Foto ©berggeist007 / pixelio.de

More available books at **www.hansebooks.com**

A TREATISE

ON THE

CONSTRUCTION AND USE

OF

MILLING MACHINES

MADE BY

BROWN & SHARPE MFG. CO.,

PROVIDENCE, R. I., U. S. A.

MANUFACTURERS OF

MACHINERY AND TOOLS.

PROVIDENCE, R. I.
BROWN & SHARPE MANUFACTURING COMPANY
1896.

PREFACE.

Our chief object in writing this book is to assist those who are not familiar with the construction or use of Universal or Plain Milling Machines. We desire to aid in having the machines well understood, properly cared for and profitably operated. Beyond this, we hope we have written somewhat of interest to those well versed in the theory and practice of milling.

CONTENTS.

FIGURES SHOWING CAPACITY OF MACHINES.

We have placed in connection with the name and immediately under the number of each machine, the figures that best indicate its capacity—the object being to assist those who desire to quickly compare machines, or wish to remember or designate them by their size in a way that is customary with lathes and planing machines.

The first figure indicates the automatic longitudinal feed of table, the second its transverse movement and the third the distance it can be lowered from the centre of spindle.

For example the No. 1 Universal Milling Machine is headed No. 1, 17 in. x 6 in. x 18 in. Universal Milling Machine, and indicates that the table has an automatic longitudinal feed of 17 in., a transverse movement of 6 in., and can be lowered 18 in. from centre of spindle.

DIMENSIONS OF MACHINES.

The Dimensions and Details of Machines given are for the machines made at the time of issuing this treatise. Changes which may be made in future lots will be indicated in the edition of the catalogue next following such change. Catalogues are mailed on application.

UNIVERSAL MILLING MACHINES.

A Universal has all the movements of a Plain Machine, and in addition, the table is fed automatically at various angles to the axis of the spindle, and the spiral head is so made and connected with the feed screw that a positive rotary movement may be given to the work.

Besides doing the work of a Plain Milling Machine it can thus be used for an almost unlimited variety of other operations in the manufacture of machinery and tools.

" Peculiarly American in its design and construction*, this machine has done more for the introduction and demand for American machinery in Europe than any other.

"As exhibited by J. R. Brown & Sharpe in 1867 at the Paris Exhibition, it attracted the attention of engineers, machine tool builders, gun and sewing machine makers, and the directors of government arsenals and other large engineering works. It was one of those special tools by the aid of which Americans have been enabled to so far improve the construction of their machinery, that they were at once successful competitors for large orders of special and other machinery for governments and other large industries. Europe from that time forward could not afford to be without the Universal Milling Machine. Those sent for exhibition were sold, soon after the opening, to Continental machine tool builders, ostensibly for the purpose of using them as copies from which to build others.

" The workmanship on these machines was of such a quality that the purchasers found difficulty in producing others of equal character, except certain parts which were reproduced on these same tools; thus showing the imperative necessity of parent tools containing within their own structure the inherent qualities of truth and the highest excellence.

" One of the minor virtues in a good machine tool is general convenience, easy access to the various parts, the arrangement of handles for manipulation, and their being so combined that the

---

*American Machinist, April, 1879, referring to No. 1 Universal Milling Machine.

attendant has not to waste his time in moving from the point of action in order to reach them. Another highly important point is that every tool should not only be sufficiently well made, but that the means of adjustment and compensation for wear should be such that the general soundness of the fitting can be maintained, and that the work produced by the machine during its life time should be perfect, and require no fitting afterwards. This is the true condition to be reached with tools, and a high standard is not secured unless the tool has a combination of all the virtues which should characterize a machine tool in this age of special machinery.

"At the Vienna Exhibition of 1873, Continental tool builders exhibited their copies of American Universal Milling Machines and Revolving Head Screw Machines, " System Brown & Sharpe," (so noted in the illustrated catalogue of one firm). Decidedly the best copies of these machines were those exhibited by the Alsatian firm of Ducommun & Steinlen of Mulhouse.

"At this exhibition, however, preference was given to the tools exhibited by Brown & Sharpe, and the consequence was that their tools were sold earlier and at a better price than the copies, some of which were very creditable.

" The late Paris Exhibition was especially rich in creditable productions of machinery of American design, very notable among which were copies of the Brown & Sharpe Universal Milling Machine.

" These were exhibited by several tool builders in the French, as also by a prominent English firm in the British section, who were considerably chagrined at the reply of the International Judges, who, in answer to their appeal for an award higher than the silver medal (of which they had received notice), informed them that with machinery as with the fine arts, 'copies were not entitled to so high an award as were the originals.'

" No gold medal awarded was more deserved than that given the Brown & Sharpe Manufacturing Co., for their fine exhibit of special machine tools, from which Continental and other firms are now constructing their copies."

" In 1889 at the Paris Exposition our exhibit was again honored. It received the " Grand Prix," the highest possible distinction.

The Universal Milling Machine, illustrated and described in the following pages, form, with the exception of the No. 4, a series of machines which was presented to the public generally for the first time at the World's Columbian Exposition in Chicago in 1893. In connection with this line of the new machines we exhibited the first No. 1 Universal Milling Machine made by us. This machine, patented by Joseph R. Brown in 1865, and extensively copied both in America and Europe, illustrated most completely the fact, that notwithstanding the number of years since its construction, there had been no changes in the fundamental principles of Universal Milling Machines.

Improvements of greater or less importance have been made in the machines from time to time, and we have embodied in their construction what we believe to be the best, making them more convenient, rigid, and better adapted for the class of work for which they are intended.

The following are some of the important features that have been applied to all of the machines of this series :

1st. **The Clamp Bed and Saddle** are held together by three bolts, far enough from the centre about which the Saddle swivels to hold the parts firmly in whatever position they may be placed, making it possible to take heavier cuts with less liability to chatter.

2nd. **The Table Feed** is at the centre, and is automatic in either direction and changed by simply operating a lever. This construction allows the table to be fed automatically when swung around to 45° either way. The Feed Stops are positive in their action.

3rd. **The Feed Connection** is with the Knee instead of the Table, which makes it possible to cut any angle of spiral within the capacity of the machines without interference of the Universal Joints, and the top is clear for long work. The Foot Stock may be removed without taking out the bolt.

4th. **The Centre in Overhanging Arm** is adjustable, and the Spindle is provided with a thread on the end, so that a face mill, chuck, etc., can be used.

5th. **The Table** is provided with a pan and channels for receiving the oil.

DIMENSIONS OF UNIVERSAL MILLING MACHINES.

NO. OF MACHINE.	1	2	3	4*
No. of Taper Hole in Spindle,	10	10	11	11
Distance from Centre of Spindle to O. H. Arm.	5 1-2''	5 1-2''	6 3-8''	7 1-4''
Greatest Distance from End of Spindle to Centre in O. H. Arm.	10 1-2''	14''	18''	24''
Back Geared.	No	Yes	Yes	Yes
Working Surface of Table.	28''x5 1-8''	32''x6 3-4''	40''x8''	48''x9 3-4''
Transverse movement of Table.	6''	6 1-2''	7 1-2''	8 1-2''
Greatest Distance from Centre of Spindle to Top of Table.	18''	17 1-2''	18 1-2''	19''
Length of Automatic Feed.	17''	20 1-2''	25''	28 1-2''
No. of changes of Feed.	6	12	16	12
Variations in Feed to one Rev. of Spindle.	.004'' to .073	.005'' to .120''	.003'' to .302''	.004'' to .214''
Index Centres Swing.	8''	10''	12''	14''
Index Centres Take.	14''	15''	21''	26''
Net Weight.	1570 lbs.	1750 lbs.	2800 lbs.	4150 lbs.
Floor Space.	61''x59''	70''x66''	84''x75''	98''x87''

*Design 1893.

No. 1

17 in. x 6 in. x 18 in.

UNIVERSAL MILLING MACHINE.

DESIGN OF 1895.

The table has an automatic longitudinal feed of 17″, a transverse movement of 6″, and can be lowered 18″ from centre of spindle.

The centres swing 8″ in diameter and take 14″ in length.

No. 1 UNIVERSAL MILLING MACHINE.

The size and capacity of the No. 1 Universal Milling Machine adapt it for use in the ordinary shop.

It is pre-eminently a tool making machine, as is indicated by the fact that in the department where we make our jigs and fixtures, there are nearly as many No. 1 Universal Milling Machines as engine lathes. In all we have about forty of these machines in operation in our shops.

The Spindle has a hole $\frac{21}{32}''$ in diameter its entire length, and at the front end a No. 10 taper hole. The spindle is ground and lapped and runs in bronze boxes that can be adjusted. The front box can be tightened by the nut A, Figure 8, which brings the shoulder of the spindle against washers in front of the frame of the machine. The rear box is adjusted by tightening the nut C; the front end of spindle is threaded, and is provided with a guard nut D, which protects the thread when not in use.

The Cone has four steps, the largest $10\frac{1}{2}''$ in diameter, for 3'' belt.

The Overhanging Arm has an adjustable centre, and can be easily reversed to receive an attachment, turned out of the way, or removed. The distance from the centre of the spindle to the arm is $5\frac{1}{2}''$; greatest distance from end of spindle to centre in arm, $10\frac{1}{2}''$

The Table, including oil channels and pans, is $31''$ long and $6\frac{1}{2}''$ wide, has a working surface $28'' \times 5\frac{1}{8}''$, and a T-slot $\frac{5}{8}''$ wide. By means of the shaft C, Figure 8, it can be moved transversely 6''. One revolution of the shaft moves the table $.2''$.

The Saddle, which carries the table, pivots in the clamp bed, and is rigidly clamped to it by three bolts, which slide in circular slots and allow the table to be set at any angle to $45°$ either way from zero.   The saddle and knee are clamped by nuts with fixed handles, thus dispensing with wrenches.

The Knee can be moved so that the top of the table at its lowest point will be 18″ from the centre of the spindle.   One revolution of crank moves the knee .1″.

The Stop Rod has sliding collars, J J, Figure 8, which can be readily set at any desired point when it is desired to limit the movement of the knee.

The Feed of table, of 17″, is automatic in either direction, and as it is central can be changed by a simple movement of the lever on front of table.   It can be used running in either direction, with the table at any angle to 45°.   The feed is driven from the feed cone through the universal joints and telescopic shafts on the side of the machine, page 5, the worm A, Figure 2, and worm wheel B, through the bevel gears, and the shaft D, the bevel gear at the upper end of the shaft which drives the loose bevel gears E E.   Motion is imparted to the table by the lever F and the clutch G.

An automatic stop is provided to release the feed at any point when running in either direction.   The auxiliary lever O, Figure 1, allows the feed to be released by hand.

A handle is provided at each end of the table for convenient operation by hand.

There are 8 changes of feed, obtained by transposing the feed cone pulleys, page 8, held in place by knurled nuts, giving a variation of feeds from .004″ to .073″ to one revolution of spindle. The lower cone pulleys are on an adjustable bracket, which allows the feed belt to be easily tightened.

Adjustable Dials graduated to read to thousandths of an inch indicate the longitudinal, transverse and vertical movements of table, and a series of graduations show in degrees the angle to the axis of the spindle at which the table is set.   The dials may be adjusted without the aid of wrench or screw driver.

The Indexing Head Stock, or the **Spiral Head** is to divide the periphery of a piece of work into a number of equal parts.   Besides this, it is so made that while the table is being moved by the feed screw a positive rotary movement may be given to the work.   The velocity ratios of these movements are

FEED TABLE FOR No. I UNIVERSAL MILLING MACHINE.

CONE ON SPINDLE								
CONE ON FEED SHAFT								
FEED PER REV. OF SPINDLE.	.004	.006	.010	.015	.021	.032	.048	.073

SPINDLE SPEEDS	TRAVEL OF TABLE IN INCHES PER MINUTE.							
68	$\frac{9}{32}$	$\frac{13}{32}$	$\frac{11}{16}$	1	$1\frac{7}{16}$	$2\frac{3}{16}$	$3\frac{1}{4}$	$4\frac{15}{16}$
110	$\frac{7}{16}$	$\frac{21}{32}$	$1\frac{1}{8}$	$1\frac{5}{8}$	$2\frac{5}{16}$	$3\frac{1}{2}$	$5\frac{1}{4}$	8
178	$\frac{23}{32}$	$1\frac{1}{16}$	$1\frac{3}{4}$	$2\frac{11}{16}$	$3\frac{3}{4}$	$5\frac{11}{16}$	$8\frac{1}{2}$	13
305	$1\frac{1}{4}$	$1\frac{13}{16}$	$3\frac{1}{16}$	$4\frac{9}{16}$	$6\frac{3}{8}$	$9\frac{3}{4}$	$14\frac{5}{8}$	$22\frac{1}{4}$

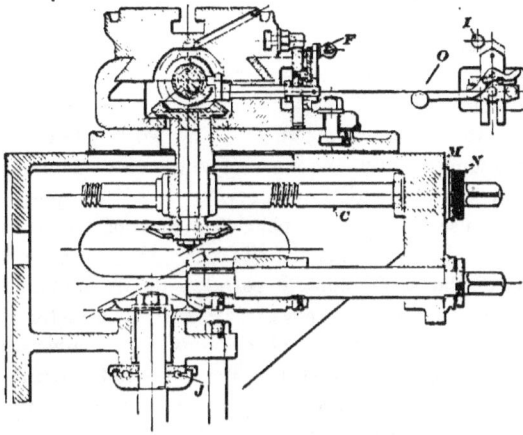

FIG. 1.

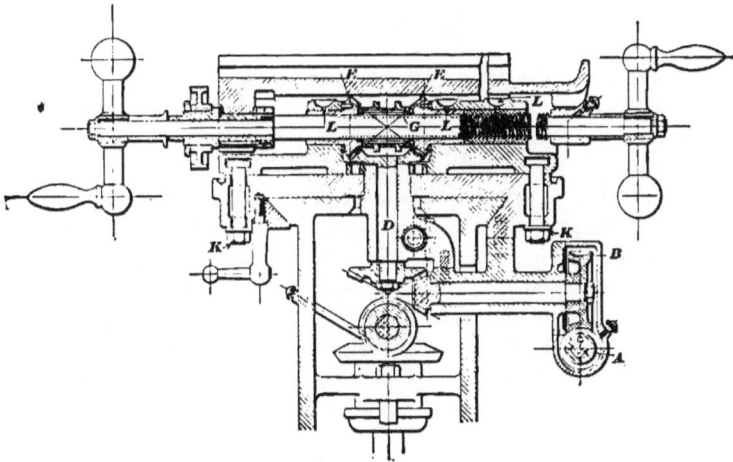

FIG. 2.

regulated by change gears, and any spiral provided for may be cut without interfering with the divisions of the work obtainable from the index plate.

Figs. 3 and 4, show the construction of the spiral head. The spiral spindle, or spiral shell, is revolved by the index crank J, through a worm and worm wheel. The worm wheel has forty teeth and consequently one turn of the index crank or worm shaft O, makes one-fortieth of a revolution of the spiral spindle, and by the use of an index plate I, a turn of the worm shaft may be divided into various definite parts and the fortieth of the revolution of the spiral spindle correspondingly sub-divided.

In connection with the index plate there is a sector which is used to obviate the necessity of counting the number of holes in the plate when dividing the work. The index crank pin P, in the crank J, may be used in any circle of holes in the index plate by adjusting the crank. The index plate can be kept from turning by the stop pin R.

The steel bushings, shown in black in the cuts, afford not only an extended bearing for the worm shaft, but also serve as a pivot for the spindle box B. The front end of the spindle can thus be raised and the spindle set to any angle from five degrees below the horizontal to a perpendicular. The angle is shown by the graduations on one side of the spiral head. The spindle receives the same arbors as the main spindle, and is threaded for a chuck. When the chuck is not in use the thread is protected by a guard nut.

Motion is transmitted to the spindle from the feed screw through change gears to two mitre gears, one of which is shown near T, Figure 3, and the other has upon its hub the gear marked "worm gear," Figure 5. By this arrangement the spindle can be automatically rotated at whatever angle it may be set.

Tables are sent with each machine, giving the change gears for cutting sixty-eight spirals. The tables call for the "gear on worm" the "first gear on stud," which means the first gear placed on stud, the "second gear on stud," or second gear placed on stud, and the "gear on screw." The spiral head and foot stock centres swing 8″ in diameter and take 14″ in length.

The chuck sent with the machine is provided with a plate

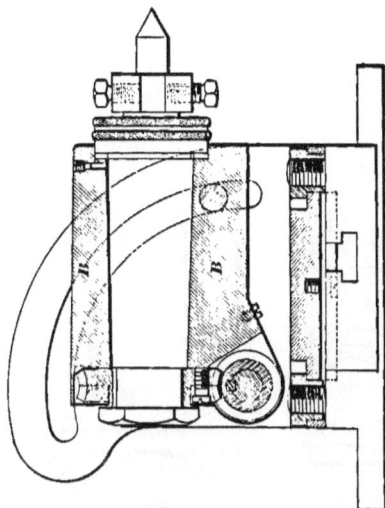

Fig. 4.

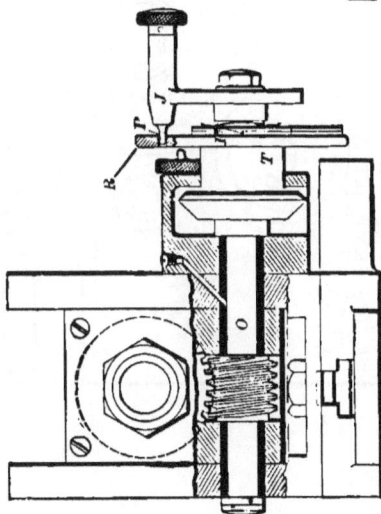

Fig. 3.

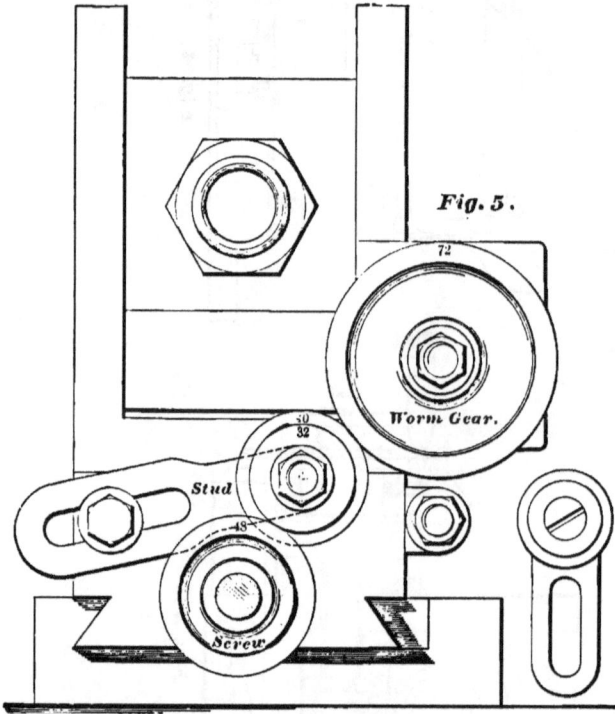

Fig. 5.

which fits the spiral head spindle.   A second chuck plate is also
sent which fits the front of main spindle.

In oiling the machine it is important not to overlook the oil
holes L L L, Figure 2.   These holes are reached from the front
of the table, when the table is placed in the three positions by
matching lines on the table with a zero line on the front of the
saddle.

Spirals not given in the table may be cut by using special
gears or by determining by calculation many combinations of
the regular gears not given in the table.

The three index plates sent with the machine have 15, 16, 17,
18, 19, 20; 21, 23, 27, 29, 31, 33, and 37, 39, 41, 43, 47, 49 holes
respectively.

By means of the raising block the spiral head may be set at any
angle on the bed.

The Vise swivels and has a graduated base, and may be set
at any angle on the bed.   The jaws are 5⅛" wide, 1⅛" deep and
will open 2¾".   The jaws are made of steel and left soft unless
otherwise ordered.   Being held by screws they may be taken off
and others designed for special work put in their places.

The Overhead Works, Figs. 6 and 7, include a shipper
rod, stops and studs for attaching the lever, also hangers with
adjustable self-oiling boxes.

The Counter Shaft has two 14 inch friction pulleys for
3½ inch belts, and is usually run at about 110 revolutions a
minute.

Floor Space measured over the extreme projections and points
of travel of the various parts, 61x59 inches.

Weight of the machine ready for shipment is about 2040
pounds.

Net Weight, about 1570 pounds.

Dimensions of box in which the machine is packed are
49x35x62 inches.

Each machine is furnished with change gears, index plates and
tables explaining the use of same, 6" 3 jawed chuck, and extra
chuck plate, vise, collet, centre rest, raising block, hand wheel,
wrenches, Treatise on Milling Machines, and everything else shown
in cut, together with overhead works.

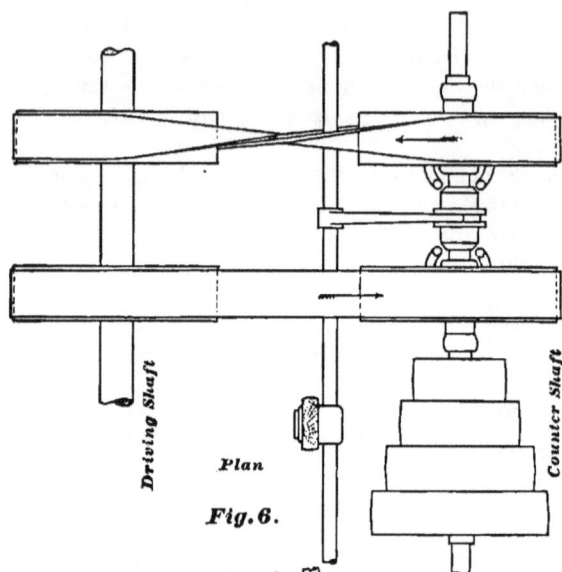

Driving Shaft

Counter Shaft

Plan

Fig. 6.

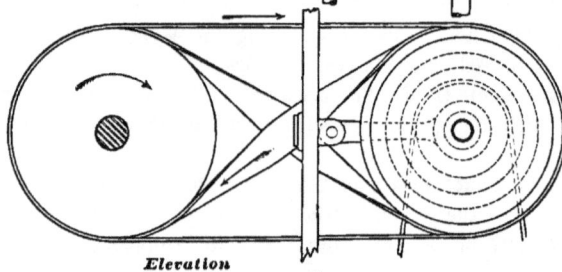

Elevation

Fig. 7.

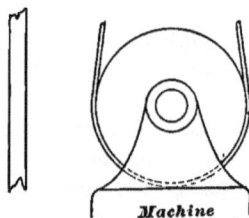

Machine

SCORING WORM GEAR

ON

No. 1 Universal Milling Machine.

No. 2

20 1-2 in. x 6 1-2 in. x 17 1-2 in.

UNIVERSAL MILLING MACHINE.

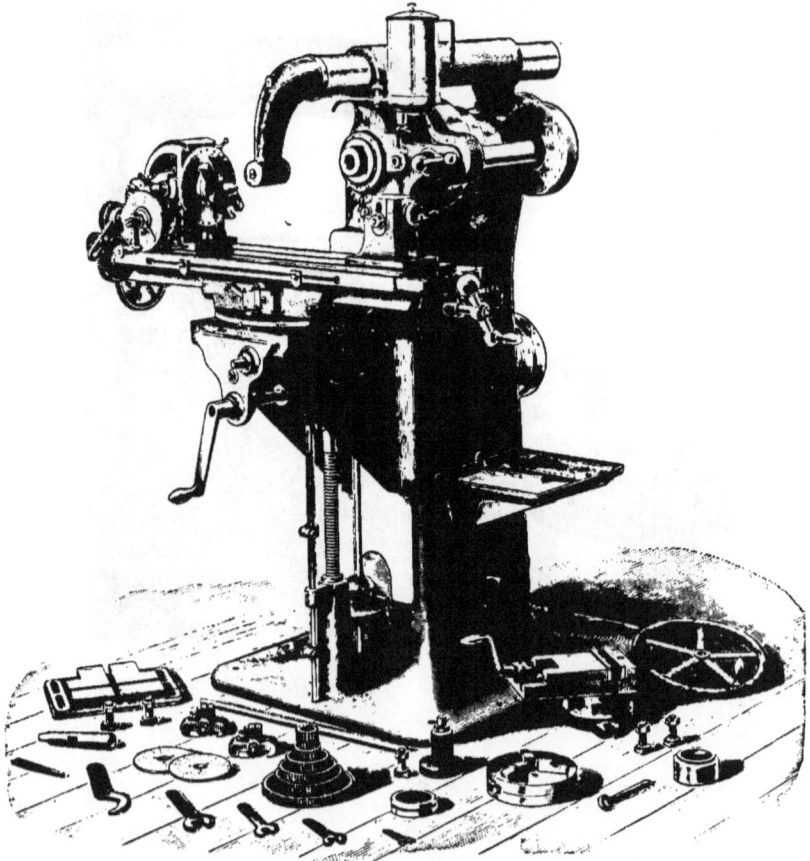

The table has an automatic longitudinal feed of 20½", a transverse movement of 6½", and can be lowered 17½" from centre of spindle.

The centres swing 10" in diameter and take 15" in length.

No. 2

20 1-2 in. x 6 1-2 in. x 17 1-2 in.

UNIVERSAL MILLING MACHINE.

Patented Feb. 5, 1884; Feb. 14, 1893; May 23, 1893.

This machine is well adapted for tool room and tool making purposes, where a more rigid, but not necessarily larger machine is required than the No. 1, Universal Milling Machine, or when saws and cutters of a comparatively large diameter are to be used.

The machine is the smallest of the Universal Milling Machines constructed with back gears, and in the main is similar to the No. 1. The principal differences, aside from the necessary changes in design by the application of back gears, are the details of the automatic feed, spiral head and foot stock. A perspective view of the machine is shown on page 16 and a cross section on page 18.

The **Spindle** has a hole $\frac{31}{32}''$ in diameter its entire length and at the front end a No. 10 taper hole. The spindle is ground and lapped and runs in bronze boxes that are adjustable. The front box can be tightened by the nut A, Fig. 8, which brings the shoulder of the collar against washers on front of the frame. The rear box is adjusted by the nut C; the front end of spindle is threaded, and is provided with a guard nut D, which protects the thread when not in use.

The **Cone** has 3 steps, the largest 10'' diameter, for 3'' belt, and is back geared, giving 6 changes of speed. The gears are covered.

. The **Overhanging Arm** has an adjustable centre, and can be easily reversed to receive an attachment, turned out of the way, or removed. The distance from the centre of the spindle to the arm is 5½''; greatest distance from end of spindle to centre in arm, 14''.

The **Table,** including oil pans and channels, is 35½'' long, 8'' wide, has a working surface 32'' x 6¾'', and 2 T slots ⅝'' wide.

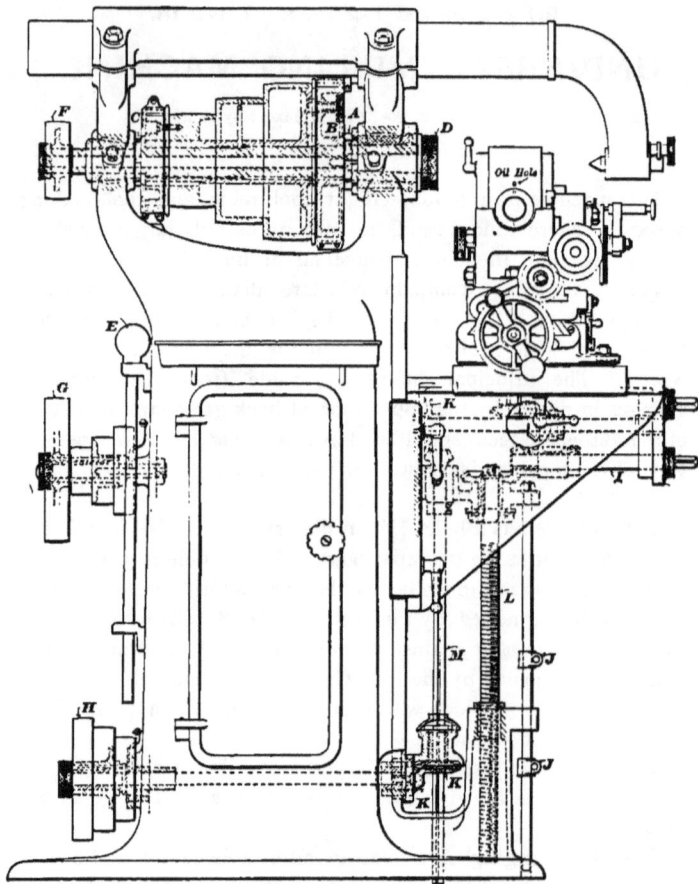

FIG. 8.

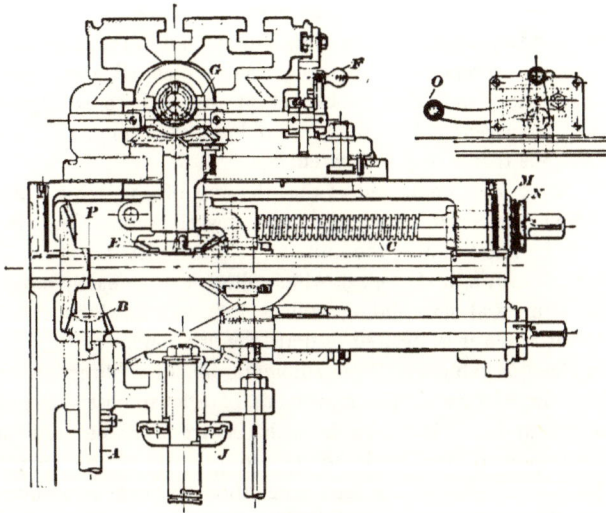

FIG. 9.

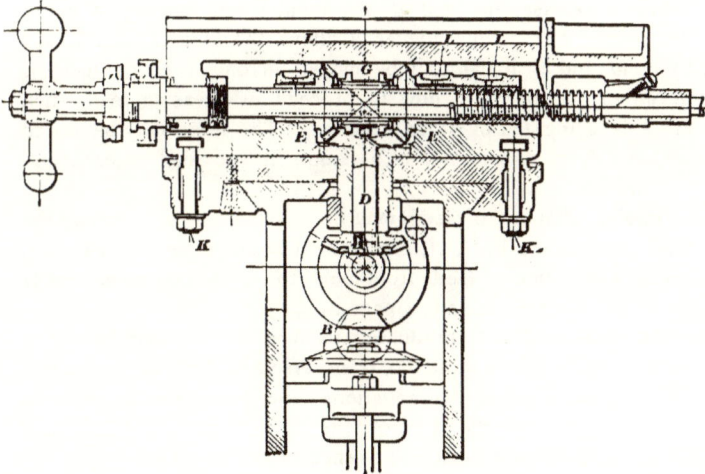

FIG. 10.

By means of the shaft C, Fig. 9, it can be moved transversely 6½". One revolution of the shaft moves the table .2".

The Saddle, which carries the table, pivots in the clamp bed, and is rigidly clamped to it by three bolts which slide in circular slots and allow the table to be set at any angle to 45 degrees, each way from zero. The saddle and knee are clamped by fixed handles, thus dispensing with wrenches.

The Knee can be moved so that the top of the table will be 17½" from the centre of spindle at its lowest point. One revolution of the crank moves the table .1". The knee rests on ball bearings as shown at J, Fig. 9, which allow it to be easily moved.

The Stop Rod has sliding collars J J, Fig. 8, which can be readily set when it is desired to limit the movement of the knee.

The Feed of table, 20½", is automatic in either direction and can be changed by a simple movement of a lever on the front of the saddle, and as it is driven from the centre it can be used with table clamped at any angle, to 45°, to the axis of the spindle. The feed is driven from the feed cones through bevel gears by the shaft A, Fig. 9, through the bevel gear B to shaft carrying bevel gear P, then through bevel gear E on lower end of vertical shaft. The clutch G is operated by lever F. An automatic stop is provided to release the feed at any point when running in either direction. The auxiliary lever O allows the feed to be released by hand, and as it can only be moved in one direction there is never any doubt as to how to proceed in releasing feed, when table is being fed in either direction. The table may be moved by hand from either end, as a handle is provided at each end for this purpose.

Twelve changes of feed may be obtained by transposing the feed cone pulleys F, G, H, Fig. 8, held in place by knurled nuts, giving a variation of feeds from .005" to .12" to one revolution of spindle. For feed table see page 21.

Adjustable Dials, graduated to read to thousandths of an inch, indicate the longitudinal, transverse and vertical movements of table, and a series of graduations show in degrees the angle to the axis of the spindle at which the table is set. The dials may be adjusted without the aid of wrench or screw driver.

The Spiral Head furnished with the Nos. 2, 3, and 4, design of 1893, Universal Milling Machines, presents special features in

FEED TABLE FOR No. 2 UNIVERSAL MILLING MACHINE.

SPINDLE SPEEDS	FEED PER REV. OF SPINDLE.											
	.005	.009	.014	.014	.018	.022	.029	.037	.046	.046	.073	.120
	TRAVEL OF TABLE IN INCHES PER MINUTE.											
32	$\frac{5}{32}$	$\frac{9}{32}$	$\frac{7}{16}$	$\frac{7}{16}$	$\frac{9}{16}$	$\frac{23}{32}$	$\frac{15}{16}$	$1\frac{3}{16}$	$1\frac{1}{2}$	$1\frac{1}{2}$	$2\frac{5}{16}$	$3\frac{7}{8}$
52	$\frac{1}{4}$	$\frac{7}{16}$	$\frac{23}{32}$	$\frac{23}{32}$	$\frac{15}{16}$	$1\frac{1}{8}$	$1\frac{1}{2}$	$1\frac{15}{16}$	$2\frac{3}{8}$	$2\frac{3}{8}$	$3\frac{13}{16}$	$6\frac{1}{4}$
85	$\frac{13}{32}$	$\frac{3}{4}$	$1\frac{3}{16}$	$1\frac{3}{16}$	$1\frac{1}{2}$	$1\frac{7}{8}$	$2\frac{7}{16}$	$3\frac{1}{8}$	$3\frac{15}{16}$	$3\frac{15}{16}$	$6\frac{1}{4}$	$10\frac{1}{4}$
126	$\frac{5}{8}$	$1\frac{1}{8}$	$1\frac{3}{4}$	$1\frac{3}{4}$	$2\frac{1}{4}$	$2\frac{3}{4}$	$3\frac{5}{8}$	$4\frac{11}{16}$	$5\frac{13}{16}$	$5\frac{13}{16}$	$9\frac{1}{4}$	$15\frac{1}{8}$
202	1	$1\frac{13}{16}$	$2\frac{13}{16}$	$2\frac{13}{16}$	$3\frac{5}{8}$	$4\frac{7}{16}$	$5\frac{7}{8}$	$7\frac{1}{2}$	$9\frac{1}{4}$	$9\frac{1}{4}$	$14\frac{3}{4}$	$24\frac{1}{4}$
330	$1\frac{5}{8}$	3	$4\frac{5}{8}$	$4\frac{5}{8}$	$5\frac{15}{16}$	$7\frac{1}{4}$	$9\frac{5}{8}$	$12\frac{1}{4}$	$15\frac{1}{8}$	$15\frac{1}{8}$	24	$39\frac{1}{2}$

SPIN.

STUD

FEED SHAFT

that the form admits of the body's being clamped as solidly in one position as in another by two bolts which are placed in a convenient position on the side, and also that when the spindle is at an angle of 90° with the bed the end of the spindle is comparatively low, thus making it very rigid in this position.   In doing many kinds of work, as in cutting the flutes in reamers, end mills, etc., the use of the worm and worm wheel can be dispensed with, and indexing done by revolving the spindle by hand.

The motion is transmitted from the feed screw through change gears to two spiral gears.   By this arrangement the spindle can be automatically rotated at whatever angle it may be set.

Tables are sent with each machine, giving the change gears for cutting sixty-eight spirals.   The tables call for the "gear on worm," the "first gear on stud," which means the first gear placed on stud, "the second gear on stud," or second gear placed on stud, and the "gear on screw."

When it is desired to rotate the spindle independently of the worm and worm wheel, the worm A, Fig. 12, is arranged so that it may be thrown out of gear with the worm wheel B, when the spindle may be turned by hand and locked by the index plate C and pin D.   An additional clamp is provided to hold the spindle, so that the strain will not come on the index pin and plate.

To throw the worm out of gear : turn the knob E (by means of the pin wrench) about a quarter of a revolution in the reverse direction to that indicated by the arrow stamped on knob E ; this will loosen nut G, which holds eccentric bushing H ; then with the fingers turn at the same time both knobs E and F, and the eccentric bushing H will be revolved and the worm disengaged from the worm wheel.

To replace the worm : turn the knobs E and F in the direction marked by the arrow, and tighten the knob E with the pin wrench.

The Foot-stock, Fig. 11, has an adjustable centre ; two taper pins, one of which is shown at Z, are used to accurately locate this centre in line with the head-stock centre.   When it is desirable to set at an angle out of parallel with the base, as in cutting taper reamers, drills, etc., the centre can be elevated or depressed by means of a rack and pinion actuated by the nut U. The centre is firmly held in any position by the nuts W, X, and Y.

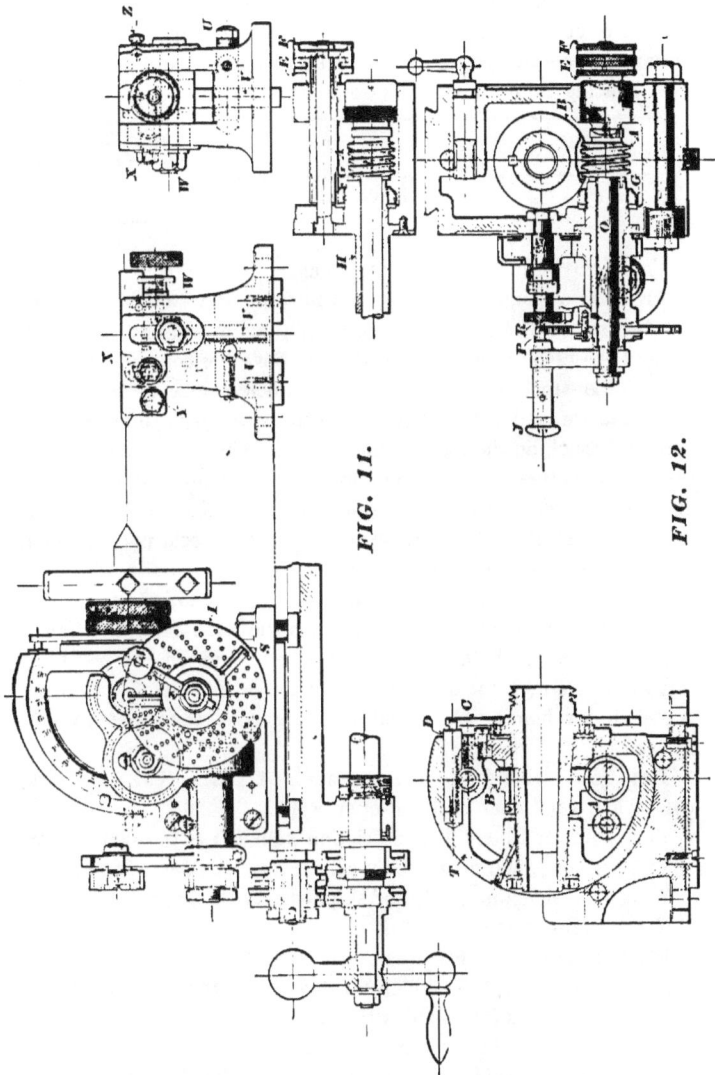

FIG. 11.

FIG. 12.

The advantage of this is easily appreciated from the fact that the use of centres that can not be adjusted work is apt to become cramped at certain positions during its revolution, and as a result even spacing cannot always be obtained.

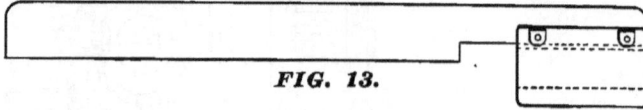

FIG. 13.

To quickly adjust the foot-stock centre in line with the spiral head centre a gauge, consisting of a bushing and a blade, Fig. 13, can be used. The bushing fits on a milling arbor, and the lower edge of the blade is the same distance above the centre as the top of the foot-stock centre. The head stock centre is set at the desired angle and the foot-stock centre is set so as to be parallel with and touching the edge of the blade. When it is correctly set the work revolves as freely as though both head and foot-stock centres were parallel with the table. If it is desirable to obtain angles greater than those allowed by the foot-stock, parallels can be used under the foot-stock.

The Frame is hollow and fitted as a closet to hold the small parts that accompany the machine. On the left side of the frame there is a pan for holding small tools, etc., and on the front of this there is a rack for wrenches. On the other side of the frame is a shelf for holding the spiral head and vise when they are not in use.

The Vise swivels and has a graduated base. The jaws are 5⅛" wide, 1⅛" deep and will open 2¾".

The Counter-Shaft has 2 friction pulleys 14" in diameter for 3½" belts, and should run about 180 revolutions per minute.

Weight of machine ready for shipment, about 2275 lbs.

Net Weight, about 1750 lbs.

Floor Space, 70"x66".

Dimensions of box in which machine is shipped, 49"x35"x62".

Each machine is furnished with change gears, index plates and tables explaining the use of same, 6" 3-jawed chuck, and extra chuck plate, vise, collet, centre rest, raising block, hand wheel, wrenches, Treatise on Milling Machines, and everything shown in cut, together with overhead works.

MILLING FACE OF UNIVERSAL MILLING MACHINE KNEE.

No. 3

25 in. x 7 1-2 in. x 18 1-2 in.

UNIVERSAL MILLING MACHINE.

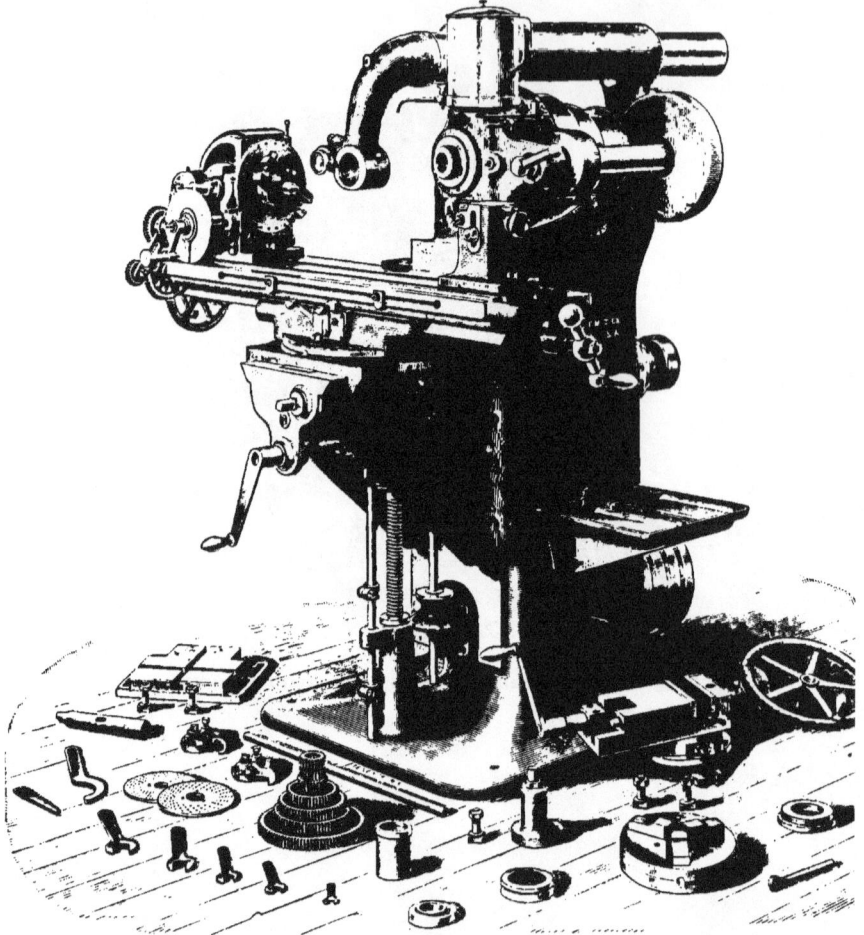

The table has an automatic longitudinal feed of 25″, a transverse movement of 7½″, and can be lowered 18½″ from centre of spindle.

The centres swing 12″ in diameter and take 21″ in length.

No. 3

25 in. x 7 1-2 in. x 18 1-2 in.

UNIVERSAL MILLING MACHINE.

With Hand or Power Transverse and Vertical Feeds.

Patented Feb. 5, 1884; Feb. 14, 1893; May 23, 1893.

The machine is similar in design to the No. 2 Universal Milling Machines and has the additional feature that it can be furnished, when desired, with Power Transverse and Vertical Feeds.

The Spindle has a hole ¾" in diameter its entire length and runs in bronze boxes provided with means of compensation for wear. For method of adjusting see page 17. It is ground and lapped. The front end is threaded and has a No. 11 taper hole.

The Cone has 3 steps, the largest 11" diameter, for 3½" belt and is back geared, giving 6 changes of speed. The gears are covered.

The Overhanging Arm has a bearing for outer end of arbor, etc., as well as an adjustable centre. It can be easily reversed to receive an attachment, turned out of the way, or removed. The distance from the centre of the spindle to the arm is 6⅜"; greatest distance from end of spindle to centre in arm, 18".

The Table, including oil pans and channels, is 44½" long, 9½" wide, has a working surface 40" x 8", 2 T slots ⅝" wide, a transverse movement of 7½", and can be lowered 18½" from centre of spindle.

The details of the **Power Transverse Feed** mechanism, furnished with this machine when desired, are shown in Figs. 14 and 15. The clutch gear A, Fig. 14, revolves constantly and is driven by the gear G, Fig. 16. The transverse or cross feed is operated by turning knob B, Figs. 14 and 15; this operating the lever C

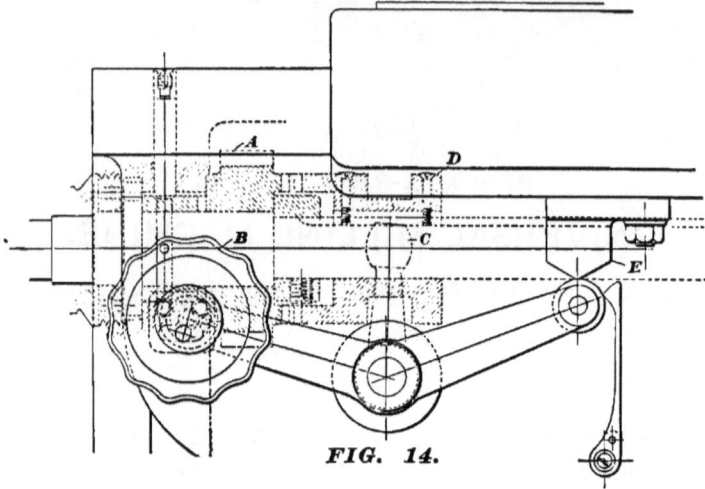

FIG. 14.

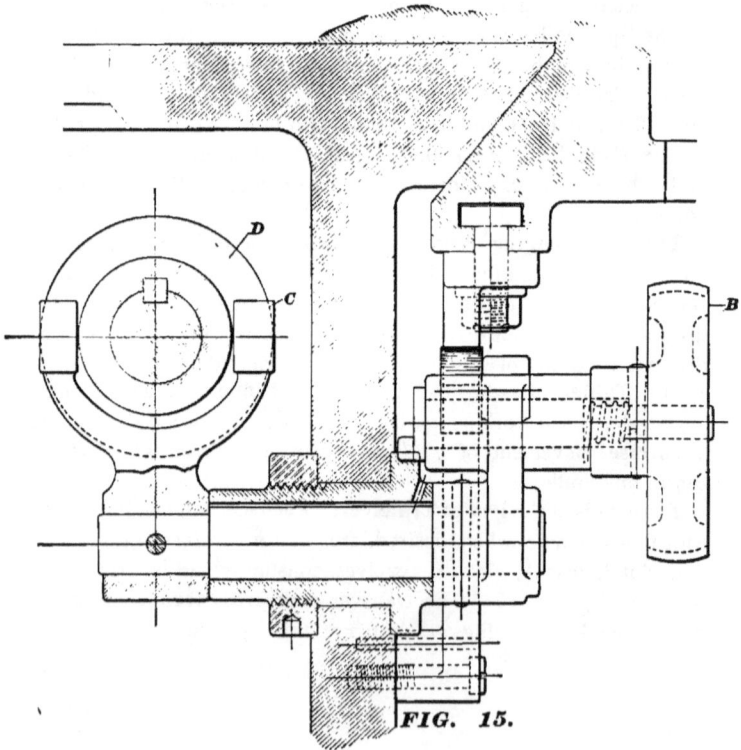

FIG. 15.

throws the clutch D into mesh with the clutch gear A, Fig. 14. The movable stop E can be set to release the feed automatically at any desired point.

The Saddle, which carries the table, pivots in the clamp bed, and is rigidly clamped to it by three bolts which slide in circular slots and allow the table to be set at any angle to 45 degrees, each way from zero. The saddle and knee are clamped by fixed handles, thus dispensing with wrenches.

The Knee can be lowered 18½" from centre of spindle, and has a stop rod with sliding collars which may be quickly set at any desired point. The details of the **Power Vertical Feed,** furnished on this machine when desired, are shown in Fig. 16. The vertical feed is driven from the feed cone pulleys through the shaft A, and bevel gears B B. The direction of the feed is determined by clutch C, which can be changed by loosening a knurled nut on the side of the box protecting the gears and clutch, and moving the clutch up or down and again tightening the nut. The driving shaft D transfers the power through the bevel gears E E to shaft F, carrying the spur gear G, meshing into the gear at the side of the clutch H ; to throw the feed in the knob I is drawn out, which engages the clutch H with the continuously running spur gears, thus driving the bevel gears K K, which drive the screw L. The feed can be released instantly by the handle J, or the trip lever M, which is actuated by the dogs, at the end of the sliding key N. The stops O O can be set at any desired point, relatively to the lug P, to determine the position at which it is desired to release this feed.

The Feed of table, 25", is automatic in either direction, and can be changed by a simple movement of a lever on the front of the saddle, and as it is driven from the centre it can be used with table clamped at any angle, to 45°, to the axis of the spindle. There are 16 changes of feed obtained by transposing the feed cone pulleys, held in place by knurled nuts, page 31, varying from .003" to .302" to one revolution of the spindle. The table may be moved by hand from either end, as at each end a handle is provided for this purpose.

Adjustable Dials, graduated to read to thousandths of an inch, indicate the longitudinal, transverse and vertical movements

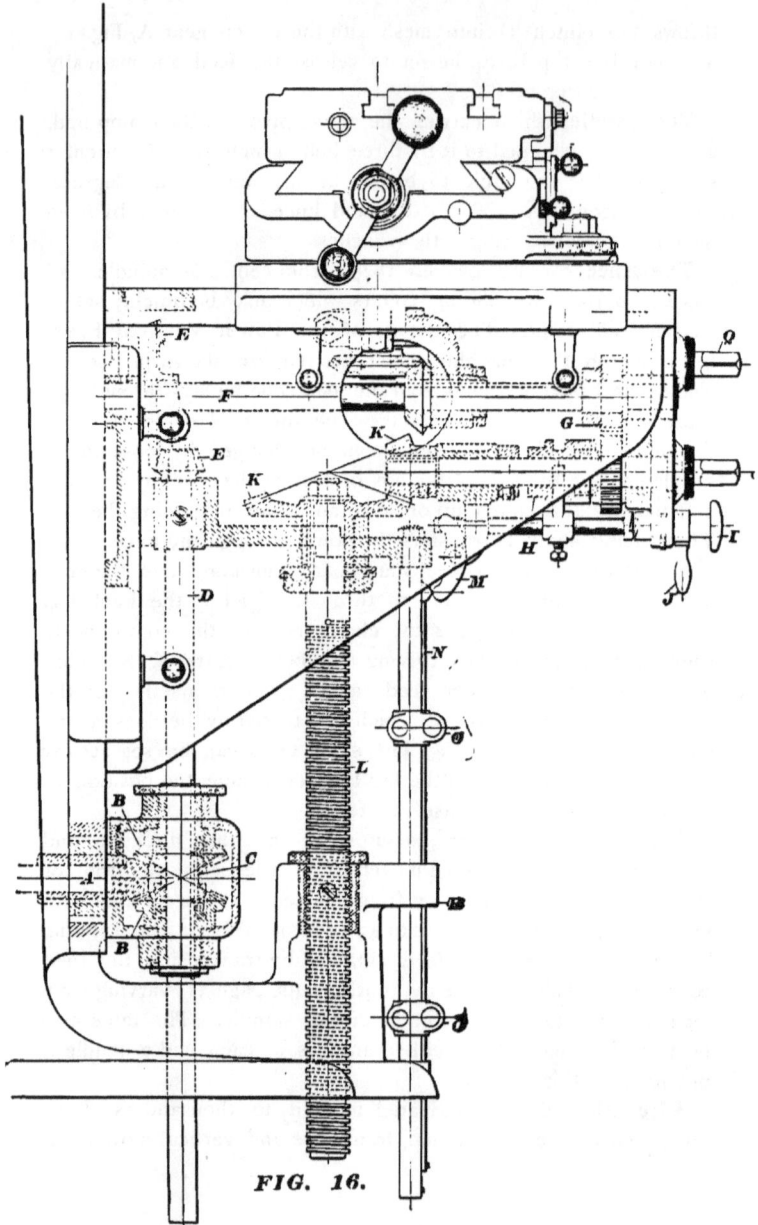

FIG. 16.

FEED TABLE FOR NO. 3 UNIVERSAL MILLING MACHINE.

SPIN																	
STUD																	
FEED SHAFT																	
FEED PER REV. OF SPINDLE	.003	.005	.008	.012	.016	.018	.024	.028	.036	.041	.057	.062	.087	.129	.193	.302
SPINDLE SPEEDS	TRAVEL OF TABLE IN INCHES PER MINUTE.															
26	$\frac{1}{16}$	$\frac{1}{8}$	$\frac{7}{32}$	$\frac{5}{16}$	$\frac{13}{32}$	$\frac{15}{32}$	$\frac{5}{8}$	$\frac{23}{32}$	$\frac{15}{16}$	$1\frac{1}{16}$	$1\frac{1}{2}$	$1\frac{5}{8}$	$2\frac{1}{4}$	$3\frac{3}{8}$	5	$7\frac{7}{8}$
39	$\frac{1}{8}$	$\frac{3}{16}$	$\frac{5}{16}$	$\frac{15}{32}$	$\frac{21}{32}$	$\frac{11}{16}$	$\frac{15}{16}$	$1\frac{1}{16}$	$1\frac{3}{8}$	$1\frac{5}{8}$	$2\frac{1}{4}$	$2\frac{7}{16}$	$3\frac{3}{8}$	5	$7\frac{1}{2}$	$11\frac{3}{4}$
60	$\frac{3}{16}$	$\frac{5}{16}$	$\frac{15}{32}$	$\frac{23}{32}$	$\frac{31}{32}$	$1\frac{1}{16}$	$1\frac{7}{16}$	$1\frac{11}{16}$	$2\frac{3}{16}$	$2\frac{7}{16}$	$3\frac{7}{16}$	$3\frac{3}{4}$	$5\frac{1}{4}$	$7\frac{1}{2}$	$11\frac{5}{8}$	$18\frac{1}{8}$
143	$\frac{7}{16}$	$\frac{23}{32}$	$1\frac{1}{8}$	$1\frac{11}{16}$	$2\frac{5}{16}$	$2\frac{9}{16}$	$3\frac{7}{16}$	4	$5\frac{1}{8}$	$5\frac{7}{8}$	$8\frac{1}{8}$	$8\frac{7}{8}$	$12\frac{1}{2}$	$18\frac{1}{2}$	$27\frac{1}{4}$	43
214	$\frac{21}{32}$	$1\frac{1}{16}$	$1\frac{11}{16}$	$2\frac{9}{16}$	$3\frac{7}{16}$	$3\frac{7}{8}$	$5\frac{1}{8}$	6	$7\frac{3}{4}$	$8\frac{3}{4}$	$12\frac{1}{4}$	$13\frac{1}{4}$	$18\frac{5}{8}$	$27\frac{1}{2}$	$41\frac{1}{2}$	$64\frac{1}{2}$
325	$\frac{31}{32}$	$1\frac{5}{8}$	$2\frac{5}{8}$	$3\frac{7}{8}$	$5\frac{3}{16}$	$5\frac{7}{8}$	$7\frac{7}{8}$	$9\frac{1}{8}$	$11\frac{3}{4}$	$13\frac{3}{8}$	$18\frac{1}{2}$	$20\frac{1}{8}$	$28\frac{1}{4}$	42	63	98

. of table, and a series of graduations show in degrees the angle to
the axis of the spindle at which the table is set.   The dials may
be adjusted without the aid of wrench or screw-driver.

The **Spiral Head** has indexing mechanism by which the
periphery of a piece of work may be divided into equal parts, and
the velocity of the rotary motion of its spindle, or of the work,
relative to the speed of the feed screw, is regulated by change
gears at the end of the table.   Any spiral, of the 68 provided for,
may be cut without interfering with the divisions obtainable from
the index plates sent with the machine.   A plate for rapid indexing of work into 24 or less divisions is placed directly on the
spindle, and the worm which turns the spindle may be thrown
quickly out of gear by a knurled knob on the back of the spiral
head to allow for this direct or plain indexing.   The spindle of
the spiral head may be moved continuously, or through any
required portion of a revolution.   The spiral head may be set at
any angle on the table by use of the raising block.   The front
end of the spindle has a No. 11 taper hole and is threaded.

The head can be set and rigidly clamped, at any angle between
10 degrees below the horizontal and 10 degrees beyond the
perpendicular.   The side of the spiral head is graduated in degrees
to show the angle of the elevation of the spindle.   The Spiral
Head and Foot-Stock Centres swing 12″ in diameter, and take
21″ in length.

The **Foot-Stock** spindle may be raised vertically, and set at an
angle in a vertical plane.   By this arrangement the spiral head
and foot-stock spindles may, in ordinary use, be kept in line
when the front of the spiral head spindle has been elevated or
depressed.

For detailed description, sketch and method of using spiral
head and foot-stock see pages 20 to 24.

The **Frame** is hollow, and fitted as a closet to hold the small
parts that accompany the machine.   On the left side of the frame
there is a pan for holding small tools, etc., and on the front of
this there is a rack for wrenches.   On the other side of the frame
is a shelf for holding the spiral head and vise when they are not
in use.

The **Vise** swivels and has a graduated base.   The jaws are
$6\frac{1}{8}$″ wide, $1\frac{7}{16}$″ deep, and will open $3\frac{5}{8}$″.

The Counter-Shaft has friction pulleys 16" in diameter for 4" belts, and should run about 175 revolutions per minute.

Weight of machine ready for shipment, about 3575 lbs.

Net Weight, about 2800 lbs.

Floor Space, 84" x 75".

Dimensions of box in which machine is shipped, 59" x 40" x 67".

Each machine is furnished with change gears, index plates, and tables explaining the use of same, 8" 3-jawed chuck, and extra chuck plate, vise, collet, centre rest, raising block, hand wheel, wrenches, Treatise on Milling Machines, and everything shown in cut, together with overhead works.

No. 4
28 1-2 in. x 8 1-2 in. x 19 in.
UNIVERSAL MILLING MACHINE.
DESIGN OF 1893

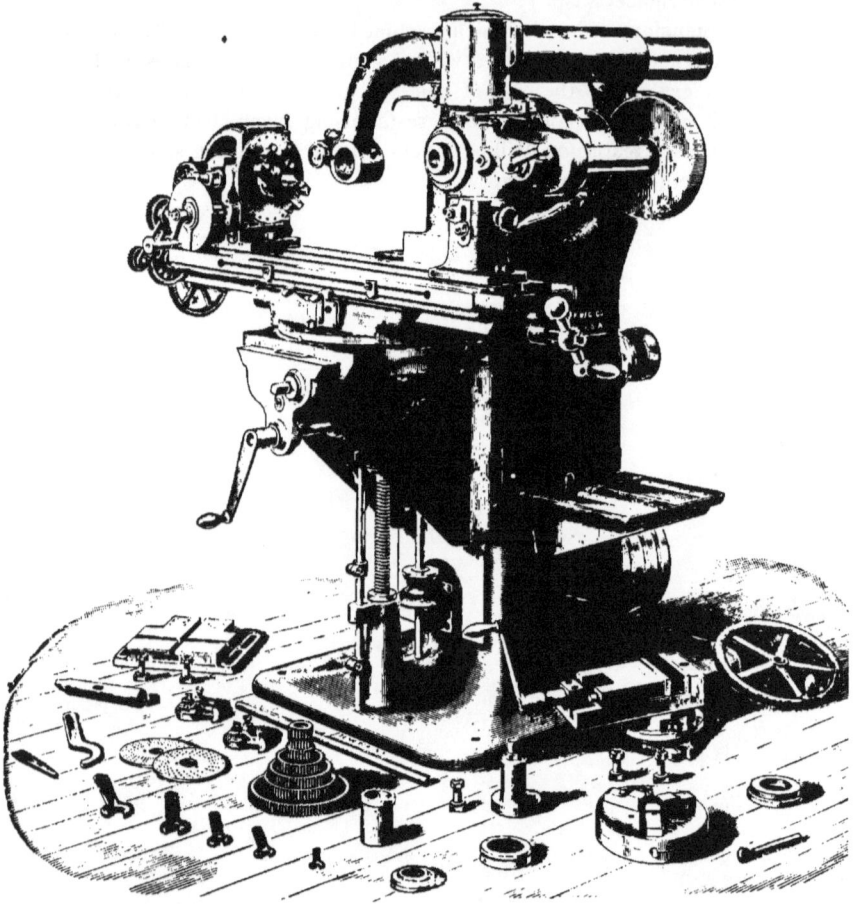

The table has an automatic longitudinal feed of 28½″, a transverse movement of 8½″, and can be lowered 19″ from centre of spindle.

The centres swing 14″ in diameter and take 26″ in length.

No. 4

28 1-2 in. x 8 1-2 in. x 19 in.

UNIVERSAL MILLING MACHINE.

With Hand or Power Transverse and Vertical Feeds.

DESIGN OF 1893.

Patented Feb. 5, 1884; Feb. 14, 1893; May 23, 1893.

This machine, while in its general construction is similar to the preceding machine described, is heavier. It is well adapted for use in shops where machine construction demands a Universal Milling Machine of this capacity.

The Spindle has a hole ¾" in diameter its entire length and runs in bronze boxes provided with means of compensation for wear. For method of adjusting see page 17. It is ground and lapped. The front end is threaded and has a No. 11 taper hole.

The Cone has 3 steps, the largest 13" in diameter, for 3½" belt and is back geared, giving 6 changes of speed. The gears are covered.

The Overhanging Arm has a bearing for outer end of arbor, etc., as well as an adjustable centre. It can be easily reversed to receive an attachment, turned out of the way, or removed. The distance from the centre of the spindle to the arm is 7¼"; greatest distance from end of spindle to centre in arm, 24".

The Table, including oil pans and channels, is 53¼" long, 11⅛" wide, has a working surface 48" x 9¾", 2 T slots ¾" wide, a transverse movement of 8½" and can be lowered 19" from centre of spindle.

The details of the **Power Transverse Feed** mechanism, furnished with this machine when desired, are similar to those on the No. 3 Universal Milling Machine described on page 27.

The Saddle, which carries the table, pivots in the clamp bed and is rigidly clamped to it by three bolts which slide in circular slots and allow the table to be set at any angle to 45 degrees, each way from zero. The saddle and knee are clamped by fixed handles, thus dispensing with wrenches.

The Knee can be lowered vertically 19″ from centre of spindle, and has a stop rod with sliding collars which may be quickly set at any desired point.

The details of the **Power Vertical Feed** mechanism, furnished with this machine when desired, are similar to those on the No. 3 Universal Milling Machine described on page 29.

The Feed of table, 28½″, is automatic in either direction and can be changed by a simple movement of a lever on the front of the saddle, and as it is driven from the centre it can be used with table clamped at any angle, to 45°, to the axis of the spindle. There are 12 changes of feed, obtained by transposing the feed cone pulleys, held in place by knurled knobs, page 37, varying from .004″ to .214″ to one revolution of spindle. A handle is provided at each end of the table for convenient operation by hand.

Adjustable Dials graduated to read to thousandths of an inch indicate the longitudinal, transverse and vertical movements of table, and a series of graduations show in degrees the angle to the axis of the spindle at which the table is set. The dials may be adjusted without the aid of wrench or screw-driver.

The Spiral Head has indexing mechanism by which the periphery of a piece of work may be divided into equal parts, and the velocity of the rotary motion of its spindle, or of the work, relative to the speed of the feed screw, is regulated by change gears at the end of the table. Any spiral of the 68 provided for may be cut without interfering with the divisions obtainable from the index plates sent with the machine. A plate for rapid index-ing of work into 24 or less divisions is placed directly on the spindle, and the worm which turns the spindle may be thrown quickly out of gear by a knurled knob on the back of the spiral head to allow for this direct or plain indexing. The spindle of the spiral head may be moved continuously, or through any required portion of a revolution. The spiral head may be set at any angle on the table by use of the raising block. The front end of spindle has a No. 11 taper hole and is threaded.

FEED TABLE FOR NO. 4 UNIVERSAL MILLING MACHINE.
DESIGN OF 1893.

SPIN.

STUD

FEED SHAFT

FEED PER REV. OF SPINDLE	.004	.006	.010	.016	.024	.025	.036	.037	.055	.090	.144	.214
SPINDLE SPEEDS	**TRAVEL OF TABLE IN INCHES PER MINUTE**											
14	$\frac{1}{16}$	$\frac{3}{32}$	$\frac{1}{8}$	$\frac{7}{32}$	$\frac{11}{32}$	$\frac{11}{32}$	$\frac{1}{2}$	$\frac{17}{32}$	$\frac{13}{16}$	$1\frac{1}{4}$	2	3
18	$\frac{1}{16}$	$\frac{3}{32}$	$\frac{3}{16}$	$\frac{7}{32}$	$\frac{7}{16}$	$\frac{7}{16}$	$\frac{21}{32}$	$\frac{11}{16}$	$1\frac{1}{16}$	$1\frac{5}{8}$	$2\frac{9}{16}$	$3\frac{7}{8}$
23	$\frac{3}{32}$	$\frac{1}{8}$	$\frac{7}{32}$	$\frac{3}{8}$	$\frac{9}{16}$	$\frac{9}{16}$	$\frac{13}{16}$	$\frac{27}{32}$	$1\frac{5}{16}$	$2\frac{1}{16}$	$3\frac{5}{16}$	$4\frac{15}{16}$
30	$\frac{1}{8}$	$\frac{3}{16}$	$\frac{5}{16}$	$\frac{15}{32}$	$\frac{3}{4}$	$\frac{3}{4}$	$1\frac{1}{16}$	$1\frac{1}{8}$	$1\frac{3}{4}$	$2\frac{11}{16}$	$4\frac{5}{16}$	$6\frac{7}{16}$
39	$\frac{5}{32}$	$\frac{7}{32}$	$\frac{3}{8}$	$\frac{5}{8}$	$\frac{15}{16}$	$\frac{15}{16}$	$1\frac{3}{8}$	$1\frac{7}{16}$	$2\frac{1}{4}$	$3\frac{1}{2}$	$5\frac{5}{8}$	$8\frac{3}{8}$
50	$\frac{3}{16}$	$\frac{5}{16}$	$\frac{1}{2}$	$\frac{13}{16}$	$1\frac{3}{16}$	$1\frac{1}{4}$	$1\frac{13}{16}$	$1\frac{7}{8}$	$2\frac{7}{8}$	$4\frac{1}{2}$	$7\frac{3}{16}$	$10\frac{3}{4}$
112	$\frac{7}{16}$	$\frac{11}{16}$	$1\frac{1}{8}$	$1\frac{13}{16}$	$2\frac{11}{16}$	$2\frac{13}{16}$	$4\frac{1}{16}$	$4\frac{1}{8}$	$6\frac{1}{2}$	$10\frac{1}{8}$	$16\frac{1}{4}$	24
144	$\frac{9}{16}$	$\frac{7}{8}$	$1\frac{7}{16}$	$2\frac{5}{16}$	$3\frac{7}{16}$	$3\frac{5}{8}$	$5\frac{3}{16}$	$5\frac{5}{16}$	$8\frac{3}{8}$	13	$20\frac{3}{4}$	$30\frac{3}{4}$
186	$\frac{3}{4}$	$1\frac{1}{8}$	$1\frac{7}{8}$	3	$4\frac{7}{16}$	$4\frac{5}{8}$	$6\frac{11}{16}$	$6\frac{7}{8}$	$10\frac{3}{4}$	$16\frac{3}{4}$	$26\frac{3}{4}$	$39\frac{3}{4}$
240	$\frac{31}{32}$	$1\frac{7}{16}$	$2\frac{3}{8}$	$3\frac{13}{16}$	$5\frac{3}{4}$	6	$8\frac{5}{8}$	$8\frac{7}{8}$	$13\frac{7}{8}$	$21\frac{1}{2}$	$34\frac{1}{2}$	$51\frac{1}{2}$
310	$1\frac{1}{4}$	$1\frac{7}{8}$	$3\frac{1}{8}$	$4\frac{15}{16}$	$7\frac{7}{16}$	$7\frac{3}{4}$	$11\frac{1}{8}$	$11\frac{1}{2}$	18	28	$44\frac{1}{2}$	$66\frac{1}{2}$
400	$1\frac{5}{8}$	$2\frac{3}{8}$	4	$6\frac{3}{8}$	$9\frac{5}{8}$	10	$14\frac{3}{8}$	$14\frac{3}{4}$	$23\frac{1}{4}$	36	$57\frac{1}{2}$	$85\frac{1}{2}$

The head can be set at any angle between 10 degrees below the horizontal and 10 degrees beyond the perpendicular, and rigidly clamped at any point. The side of the spiral head is graduated in degrees to show the angle of the elevation of the spindle. The Spiral Head and Foot-Stock Centres swing 14″ in diameter and take 26″ in length.

The Foot-Stock spindle may be raised vertically and set at an angle in a vertical plane. By this arrangement the spiral head and foot-stock spindles may, in ordinary use, be kept in line when the front of the spiral head spindle has been elevated or depressed.

For detailed description, sketch and method of using the spiral head and foot-stock see pages 20 to 24.

The Frame is hollow and fitted as a closet to hold the small parts that accompany the machine. On the left side of the frame there is a pan for holding small tools, etc., and on the front of this there is a rack for wrenches. On the other side of the frame is a shelf for holding the spiral head and vise when they are not in use.

The Vise swivels and has a graduated base. The jaws are 6⅛″ wide, 1$\frac{7}{16}$″ deep and will open 3⅝″.

The Counter-Shaft has tight and loose pulleys 16″ in diameter for 4″ belts, and should run about 175 revolutions per minute.

Weight of machine ready for shipment, about 4950 lbs.

Net Weight, about 4150 lbs.

Floor Space, 98″ x 87″.

Dimensions of box in which machine is shipped, 67″ x 46″ x 70″.

Each machine is furnished with change gears, index plates and tables explaining the use of same, 9″ 3-jawed chuck and extra chuck plate, vise, collet, centre rest, raising block, hand wheel, wrenches, Treatise on Milling Machines, and everything shown in cut, together with overhead works.

UNIVERSAL MILLING MACHINE IN OPERATION.

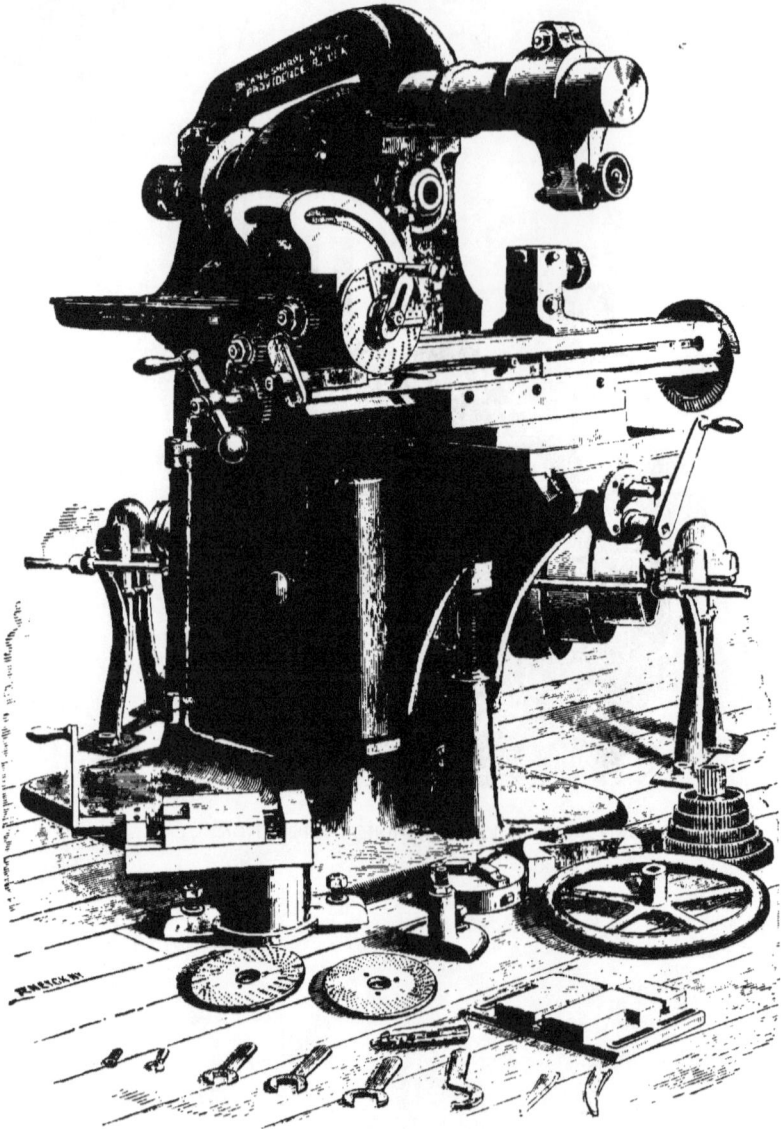

No. 4 Universal Milling Machine.
(No. 3, Prior to 1893.)

No. 4

22 in. x 6 3-8 in. x 21 in.

UNIVERSAL MILLING MACHINE.

(No. 3, Prior to 1893.)

This machine is similar in its general design to the No. 1 Universal Milling Machine made prior to 1895, but is heavier and back geared.

The Spindle runs in bronze boxes and can be adjusted to compensate for wear by tightening the nuts C and V, Fig. 17. The end thrust of the spindle is taken by a washer as shown at X. The front end has a No. 11 taper hole.

The Cone has 3 steps for $3\frac{1}{2}''$ belt and is back geared, the back gears are placed beneath the cone and may be taken out by unscrewing the screw G, taking off the lever and taking out the screw M; the lever is then put back and the eccentric shaft with the bushing pulled out.

The Overhanging Arm has a movable arbor support with an adjustable centre. Distance from centre of spindle to arm, 4".

The Table is 40" long and 7" wide, has a T slot $\frac{3}{4}''$ wide, a transverse movement of $6\frac{3}{8}''$, and can be lowered 21" from centre of spindle. It can be set at any angle to 45 degrees.

The Feed of table, of 22", is automatic in either direction. There are 6 changes of feed varying from .004" to .05" to one revolution of spindle.

Adjustable Dials, graduated to read to thousandths of an inch, indicate the transverse and vertical movements of table.

The Spiral Head and Foot-Stock Centres, similar in design to those described on page 7, swing $11\frac{1}{4}''$ in diameter and take $22\frac{3}{4}''$ in length. The head can be set at any angle from 5 degrees below the horizontal to the perpendicular. The front end

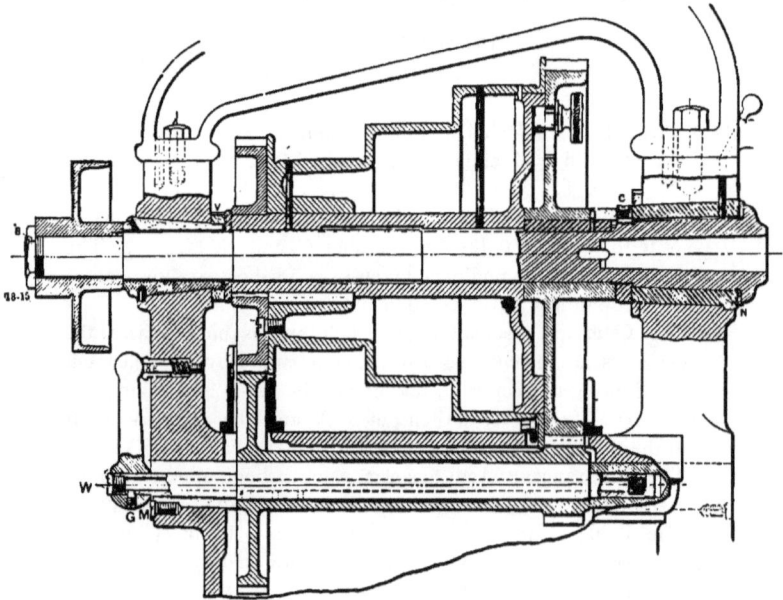

FIG. 17.

of spindle is threaded and has a No. 11 taper hole. The straight hole through spindle, at end of taper, is $1\frac{1}{4}''$ in diameter. The head can be set at any angle on table by means of the raising block.

The Vise swivels and has a graduated base. The jaws are $6\frac{1}{8}''$ wide, $1\frac{7}{16}''$ deep and will open $3\frac{5}{8}''$.

The Counter-Shaft has 2 friction pulleys $16''$ in diameter for $4''$ belts and should run about 105 revolutions per minute.

Weight of machine ready for shipment, about 3900 lbs.

Net Weight, about 3165 lbs.

Floor Space, $74'' \times 57''$.

Dimensions of box in which machine is shipped, $63'' \times 40'' \times 71''$.

Each machine is furnished with change gears, index plates, and tables explaining use of same, $9''$ 3-jawed chuck, vise, collet, centre rest, raising block, hand wheel, wrenches, Treatise on Milling Machines, and everything shown in cut, together with overhead works.

MILLING ELECTRIC MOTOR GEARS ON A PLAIN MILLING MACHINE.

PLAIN MILLING MACHINES.

The majority of milling is plain milling. Cutting spirals, or work that is milled when the table is at an angle, other than a right angle, with the spindle, form a comparatively small part of the work done even on Universal Milling Machines. Where a shop can have but one machine, without doubt it should be a Universal, for then the workman is prepared for any operation that may be required. When a second machine, however, is to be purchased the advantages of obtaining a plain milling machine should be considered. The plain machines are more simple and consequently comparatively low priced, and as the table is not swiveled it can rest directly upon the knee or other support, rendering the machine stiffer and enabling a heavier cut to be taken and frequently more work produced in any given time.

To operate a Plain Machine requires so little attention, comparatively, that frequently the man who runs it can also run a Universal Machine. On many classes of work boys can be employed, and in a number of cases one boy can operate as many as four machines. As the advantage of plain milling is more adequately appreciated the number of machines in use is rapidly increasing. At one time it was thought that only irregular shaped or small pieces could be milled advantageously, but it is now quite well known that very many plain surfaces and large pieces can be milled more profitably than they can be planed. The cutter moving constantly through the work will accomplish more than the planing tool with its intermittent action. It is also now more fully appreciated than formerly, that usually cutters can be made quite cheaply, and with care will last a long time, also that in manufacturing, even though the special cutters and fixtures are comparatively expensive, the saving in time and the superior accuracy of the work make their use profitable.

For a number of years we have used and built Plain Milling Machines, and they have formed parts of the exhibits which have been honored by the juries of award in the various world expositions during the past twenty-five years.

DIMENSIONS OF PLAIN MILLING MACHINES.

NO. OF MACHINE.	0	1	2	3	4	5
No. Prior to 1893.		4			6	8
No. of Taper Hole in Spindle.	9	10	10	11	11	12
Distance from Centre of Spindle to O. H. Arm.	5 1-8"	5 1-2"	5 1-2"	6 3-8"	7 1-4"	8 1-8"
Greatest Distance from End of Spindle to Centre in O. H. Arm.	10 1-2"	16"	16"	21"	26 1-2"	24"
Back Geared.	No	No	Yes	Yes	Yes	Yes
Working Surface of Table.	20" x 6"	32" x 7 1-2"	34" x 10"	42" x12	48" x14	54" x16"
Transverse Movement of Table.	4 1-4"	6 1-2"	6 1-2"	7	8 1-4"	9 3-4"
Transverse Movement of Table fitted with Oil Pump.				6"	7 1-8"	8 1-4"
Greatest Distance from Centre of Spindle to Top of Table.	15"	18 1-2"	18 1-2"	19 3-4"	22"	19 1-2"
Length of Automatic Feed.	16"	24"	28"	34"	42"	48"
No. of Changes of Feed.	8	8	12	16	12	8
Variations in Feed to one rev. of Spindle	.005" to .107"	.007" to .118"	.006" to .133"	.004" to .332"	.005" to .236"	.023" to .25"
Net Weight.	850 lbs.	1550 lbs.	1750 lbs.	2750 lbs.	3590 lbs.	5200 lbs.
Floor Space.	49" x 36"	69" x 46"	68" x 45"	84" x51	102" x 50"	115" x 69"

DIMENSIONS OF PLAIN MILLING MACHINES.

NO. OF MACHINE.	12	13	23	24
No. prior to 1893.	2	3	5	7
No. of Taper Hole in Spindle.	10	10	11	12
Distance from Centre of Spindle to O. H. Arm.	3 3-16''	2 7-8''	5 1-2''	5 9-16''
Greatest Distance from End of Spindle to Centre in O. H. Arm.	9 3-4''	11''	15'	18''
Working Surface of Table.	19''x 6''	27'' x 8''	50 5-8'' x 10'	60'' x 14 1-2''
Transverse Movement of Table.		3''	9'	12''
Greatest Distance from Centre of Spindle to Top of Table.	7 1-2''	6 3-8''	19''	17 5-8''
Length of Automatic Feed.	17 1-2''	15''	49''	60''
No. of Changes of Feed.	3	3		
Variations in Feed to one rev. of Spindle.	.014'' to .05	.015'' to .04''	0 to .10''	0 to .08''
Net Weight.	1600 lbs.	2240 lbs.	2600 lbs.	3700 lbs.
Floor Space.	46'' x 45''	47' x 47''	45'' x 110''	52'' x 129''

No. 0

16 in. x 4 1-4 in. x 15 in.

PLAIN MILLING MACHINE.

The table has an automatic longitudinal feed of 16″, a transverse movement of 4¼″, and can be lowered 15″ from centre of spindle.

No. O

16 in. x 4 1-4 in. x 15 in.

PLAIN MILLING MACHINE.

With Rack or Screw Feed.

Screw Feed Machine Patented May 23, 1893.

The Cut Illustrates the Rack Feed Machine.

This machine is made with either Screw or Rack Feed, and accordingly has special advantages for tool work or for manufacturing purposes. With Screw Feed the machine is more convenient for a tool maker or other operator who wishes to work exactly to a line. The table can be fed by hand to a given point with steadiness and certainty. It is always under absolute control and is practically locked when the feed is disengaged.

For manufacturing purposes, on the other hand, the machine with Rack Feed is superior, as the position of the table may be much more readily and rapidly changed by hand, and consequently time can be saved in various ways, notably, in moving the table for taking work out and placing it in position for the cut. In manufacturing, the power feed is used and the movement of the table is steady and uniform.

The Spindle has a hole $\frac{17}{32}''$ in diameter its entire length, and runs in bronze boxes provided with means of compensation for wear. For method of adjusting see page 17. The front end is threaded and has a No. 9 taper hole.

The Cone has 4 steps, the largest 10'' diameter, for $2\frac{1}{4}''$ belt.

The Overhanging Arm can be easily reversed to receive an attachment, turned out of the way, or removed. The distance from the centre of the spindle to the arm is $5\frac{5}{8}''$; greatest distance from end of spindle to centre in arm, $10\frac{1}{2}''$.

FEED TABLE FOR No. 0 PLAIN MILLING MACHINE.
RACK OR SCREW FEED.

CONE ON SPINDLE								
CONE ON FEED SHAFT								
FEED PER REV. OF SPINDLE.	.005	.008	.012	.018	.030	.044	.069	.107
SPINDLE SPEEDS	TRAVEL OF TABLE IN INCHES PER MINUTE.							
90	$\frac{7}{16}$	$\frac{23}{32}$	$1\frac{1}{16}$	$1\frac{5}{8}$	$2\frac{11}{16}$	$3\frac{15}{16}$	$6\frac{3}{16}$	$9\frac{3}{4}$
144	$\frac{23}{32}$	$1\frac{1}{8}$	$1\frac{3}{4}$	$\cdot 2\frac{9}{16}$	$4\frac{5}{16}$	$6\frac{5}{16}$	$9\frac{7}{8}$	15
225	$1\frac{1}{8}$	$1\frac{13}{16}$	$2\frac{11}{16}$	$4\frac{1}{16}$	$6\frac{3}{4}$	10	$15\frac{1}{2}$	24
360	$1\frac{13}{16}$	$2\frac{7}{8}$	$4\frac{5}{16}$	$6\frac{1}{2}$	$10\frac{3}{4}$	$15\frac{7}{8}$	$24\frac{3}{4}$	$38\frac{1}{2}$

The Table, including oil pans and channels, is 27" long, 8" wide, has a working surface 20" x 6", 3 'T' slots ½" wide, a transverse movement of 4¼", and can be lowered 15" from centre of spindle.

The Feed of table, 16", is automatic in either direction, and there are 8 changes of feed, obtained by transposing the feed pulleys, page 50, held in place by knurled nuts, varying from .005" to .107" to one revolution of spindle. The cone pulley bracket is adjustable, which allows the feed belt to be tightened.

Adjustable Dials, graduated to read to thousandths of an inch, indicate the transverse and vertical movements of table, and these dials may be adjusted without the aid of wrench or screw-driver.

The Frame is hollow and fitted as a closet to hold the small parts that accompany the machine. On the left side of the frame there is a pan for holding small tools, etc., and on the front of this there is a rack for wrenches.

The Vise, with Rack Feed Machine, is flanged and has jaws 4¼" wide, $\frac{13}{16}$" deep and will open 2".

The Vise with Screw Feed Machine swivels, and has a graduated base. The jaws are 5¼" wide, 1¼" deep and will open 2¾".

The Counter-Shaft has tight and loose pulleys, 12" in diameter, for 2½" belts, and should run about 180 revolutions per minute.

Weight of machine ready for shipment, about 1150 lbs.

Net Weight, about 850 lbs.

Floor Space, 50" x 36".

Dimensions of box in which machine is shipped, 39" x 30" x 59".

Each machine is furnished with vise, oil can, collet, wrenches, Treatise on Milling Machines, and everything shown in cut, together with overhead works.

No. 1

24 in. x 6 1-2 in. x 18 1-2 in.

PLAIN MILLING MACHINE.

(No. 4, Prior to 1893.)

The table has an automatic longitudinal feed of 24″, a trans·
verse movement of 6½″, and can be lowered 18½″ from centre
of spindle.

No. 1

24 in. x 6 1-2 in. x 18 1-2 in.

PLAIN MILLING MACHINE.

(No. 4, Prior to 1893.)

With Rack or Screw Feed.

Screw Feed Machine Patented May 23, 1893.

The Cut Illustrates the Screw Feed Machine.

This machine, when fitted with screw feed, is well adapted for tool room and jobbing use and when fitted with rack feed is well adapted for manufacturing purposes, as the table can be quickly handled for putting on and taking off work.

This machine will do heavier work than the universal machine of the same size, as the table rests directly upon the knee and the arm supports steady the cutter arbor.

The Spindle has a hole $\frac{21}{32}''$ in diameter its entire length, and runs in bronze boxes provided with means of compensation for wear. For method of adjusting see page 17. The front end is threaded, and has a No. 10 taper hole.

The Cone has 4 steps, the largest $10\frac{1}{2}''$ diameter, for $3''$ belt.

The Overhanging Arm has a hole for a centre or for a bearing for outer end of arbor, etc. It can be easily reversed to receive an attachment, turned out of the way, or removed. The distance from the centre of the spindle to the arm is $5\frac{1}{2}''$; greatest distance from end of spindle to centre in arm, $16''$. An arm support is furnished and with this in position milling can be done to $13\frac{1}{2}''$ from face of column.

The Table, including oil pans and channels, is $38''$ long, $10''$ wide, has a working surface $32'' \times 7\frac{1}{2}''$, 3 T slots $\frac{5}{8}''$ wide, a

FEED TABLE FOR No.1 PLAIN MILLING MACHINE.
SCREW FEED.

	.005	.008	.012	.018	.026	.039	.058	.089
CONE ON SPINDLE								
CONE ON FEED SHAFT								
FEED PER REV. OF SPINDLE.	.005	.008	.012	.018	.026	.039	.058	.089
SPINDLE SPEEDS	TRAVEL OF TABLE IN INCHES PER MINUTE.							
68	$\frac{11}{32}$	$\frac{17}{32}$	$\frac{13}{16}$	$1\frac{1}{4}$	$1\frac{3}{4}$	$2\frac{5}{8}$	$3\frac{15}{16}$	$6\frac{1}{16}$
110	$\frac{9}{16}$	$\frac{7}{8}$	$1\frac{5}{16}$	2	$2\frac{7}{8}$	$4\frac{5}{16}$	$6\frac{3}{8}$	$9\frac{3}{4}$
178	$\frac{7}{8}$	$1\frac{7}{16}$	$2\frac{1}{8}$	$3\frac{3}{16}$	$4\frac{5}{8}$	$6\frac{15}{16}$	$10\frac{3}{8}$	$15\frac{7}{8}$
306	$1\frac{1}{2}$	$2\frac{7}{16}$	$3\frac{11}{16}$	$5\frac{1}{2}$	$7\frac{15}{16}$	$11\frac{7}{8}$	$17\frac{3}{4}$	$27\frac{1}{4}$

FEED TABLE FOR No. 1 PLAIN MILLING MACHINE
RACK FEED.

CONE ON SPINDLE / CONE ON FEED SHAFT							
FEED PER REV. OF SPINDLE.							
.007	.010	.016	.023	.034	.051	.077	.118

SPINDLE SPEEDS	TRAVEL OF TABLE IN INCHES PER MINUTE.							
68	$\frac{15}{32}$	$\frac{11}{16}$	$1\frac{1}{16}$	$1\frac{9}{16}$	$2\frac{5}{16}$	$3\frac{7}{16}$	$5\frac{1}{4}$	8
110	$\frac{25}{32}$	$1\frac{1}{8}$	$1\frac{3}{4}$	$2\frac{1}{2}$	$3\frac{3}{4}$	$5\frac{5}{8}$	$8\frac{1}{2}$	13
178	$1\frac{1}{4}$	$1\frac{3}{4}$	$2\frac{7}{8}$	$4\frac{1}{8}$	$6\frac{1}{16}$	$9\frac{1}{8}$	$13\frac{3}{4}$	21
306	$2\frac{1}{8}$	$3\frac{1}{16}$	$4\frac{7}{8}$	$7\frac{1}{16}$	$10\frac{3}{8}$	$15\frac{5}{8}$	$23\frac{1}{2}$	36

transverse movement of 6½", and can be lowered 18½" from centre of spindle.

The Feed of table, 24", is automatic in either direction, and there are 8 changes of feed, obtained by transposing the feed pulleys, pages 54 and 55, held in place by knurled nuts, varying from .007" to .118" in the rack feed machine and from .005" to .089" in the screw feed machine, to one revolution of spindle.

Adjustable Dials, graduated to read to thousandths of an inch, indicate the transverse and vertical movements of table, and these dials may be adjusted without the aid of wrench or screw-driver.

The Frame is hollow and fitted as a closet to hold the small parts that accompany the machine. On the left side there is a pan for holding small tools, etc., and on the front of this there is a rack for wrenches.

The Vise, with Rack Feed Machine, is flanged and has jaws 5¼" wide, 1⅛" deep and opens 2¾".

The Vise, with Screw Feed Machine, swivels and has a graduated base. The jaws are 5⅛" wide, 1⅛" deep, and will open 2¾".

The Counter-Shaft has tight and loose pulleys, 14" in diameter, for 3½" belts, and should run about 110 revolutions per minute.

Weight of machine ready for shipment, about 2025 lbs.

Net Weight, about 1550 lbs.

Floor Space, 70" x 45".

Dimensions of box in which machine is shipped, 46" x 59" x 33".

Each machine is furnished with vise, oil can, collet, wrenches, Treatise on Milling Machines, and everything shown in cut, together with overhead works.

MILLING FACE AND SIDES AT ONE CUT.

No. 2

28 in. x 6 1-2 in. x 18 1-2 in.

PLAIN MILLING MACHINE.

With Rack or Screw Feed.

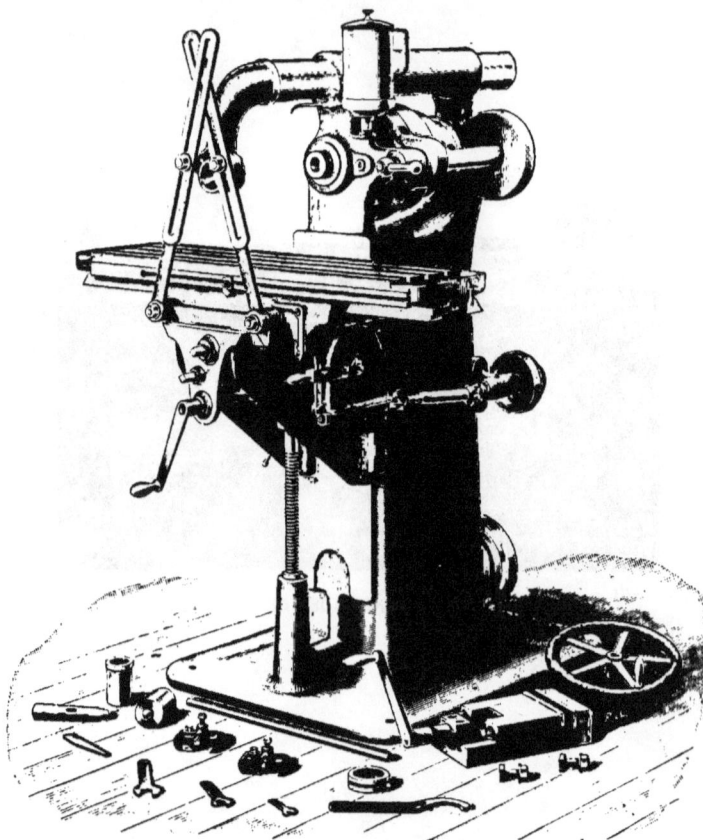

The table has an automatic longitudinal feed of 28″, a transverse movement of 6½″, and can be lowered 18½″ from centre of spindle.

No. 2

28 in. x 6 1-2 in. x 18 1-2 in.

PLAIN MILLING MACHINE.

WITH RACK OR SCREW FEED.

Screw Feed Machine Patented May 23, 1893.

The Cut Illustrates the Screw Feed Machine.

The Spindle has a hole $\frac{2\frac{1}{2}}{3\frac{1}{2}}''$ in diameter its entire length, and runs in bronze boxes provided with means of compensation for wear. For method of adjusting see page 17. The front end is threaded, and has a No. 10 taper hole.

The Cone has 3 steps, the largest 10″ diameter, for 3″ belt, and is back geared, giving six changes of speed. The gears are covered.

The Overhanging Arm has a hole for a centre, or for a bearing for outer end of arbor, etc. It can be easily reversed to receive an attachment, turned out of the way, or removed. The distance from the centre of the spindle to the arm is 5½″; greatest distance from end of spindle to centre in arm, is 16″. An arm support is furnished and with this in position milling can be done to 13½″ from face of column.

The Table, including oil pans and channels, is 40″ long, 41½″ on screw feed machine, 10″ wide, has a working surface 34″ x 10″, 3 T slots, ⅝″ wide, a transverse movement of 6½″, and can be lowered 18½″ from centre of spindle.

The Feed of table, 28″, is automatic in either direction, and there are 12 changes of feed, obtained by transposing the feed pulleys, page 60 and 61, held in place by knurled nuts, varying from .006″ to .133″ in the rack feed machine and from .005″ to .117″ in the screw feed machine, to one revolution of spindle.

FEED TABLE FOR NO. 2 PLAIN MILLING MACHINE.
SCREW FEED.

FEED PER REV. OF SPINDLE	.005	.009	.014	.014	.017	.022	.029	.036	.045	.045	.071	.117
SPINDLE SPEEDS	TRAVEL OF TABLE IN INCHES PER MINUTE.											
32	$\frac{5}{32}$	$\frac{9}{32}$	$\frac{7}{16}$	$\frac{7}{16}$	$\frac{9}{16}$	$\frac{11}{16}$	$\frac{15}{16}$	$1\frac{1}{8}$	$1\frac{7}{16}$	$1\frac{7}{16}$	$2\frac{1}{4}$	$3\frac{3}{4}$
52	$\frac{1}{4}$	$\frac{7}{16}$	$\frac{23}{32}$	$\frac{23}{32}$	$\frac{7}{8}$	$1\frac{1}{8}$	$1\frac{1}{2}$	$1\frac{7}{8}$	$2\frac{5}{16}$	$2\frac{5}{16}$	$3\frac{11}{16}$	$6\frac{1}{16}$
85	$\frac{7}{16}$	$\frac{3}{4}$	$1\frac{3}{16}$	$1\frac{3}{16}$	$1\frac{7}{16}$	$1\frac{7}{8}$	$2\frac{7}{16}$	$3\frac{1}{16}$	$3\frac{13}{16}$	$3\frac{13}{16}$	$6\frac{1}{16}$	10
126	$\frac{5}{8}$	$1\frac{1}{8}$	$1\frac{3}{4}$	$1\frac{3}{4}$	$2\frac{1}{8}$	$2\frac{3}{4}$	$3\frac{5}{8}$	$4\frac{9}{16}$	$5\frac{11}{16}$	$5\frac{11}{16}$	9	$14\frac{9}{16}$
203	1	$1\frac{13}{16}$	$2\frac{13}{16}$	$2\frac{13}{16}$	$3\frac{7}{16}$	$4\frac{7}{16}$	$5\frac{7}{8}$	$7\frac{5}{16}$	$9\frac{1}{8}$	$9\frac{1}{8}$	$14\frac{3}{8}$	$23\frac{3}{4}$
330	$1\frac{5}{8}$	3	$4\frac{5}{8}$	$4\frac{5}{8}$	$5\frac{5}{8}$	$7\frac{1}{4}$	$9\frac{5}{8}$	$11\frac{7}{8}$	$14\frac{7}{8}$	$14\frac{7}{8}$	$23\frac{1}{2}$	$38\frac{1}{2}$

FEED TABLE FOR No. 2 PLAIN MILLING MACHINE.
RACK FEED.

SPIN.

FEED SHAFT

STUD

FEED PER REV. OF SPINDLE.	.006	.010	.016	.016	.020	.025	.032	.041	.051	.051	.080	.133
SPINDLE SPEEDS	**TRAVEL OF TABLE IN INCHES PER MINUTE.**											
32	$\frac{3}{16}$	$\frac{5}{16}$	$\frac{1}{2}$	$\frac{1}{2}$	$\frac{5}{8}$	$\frac{13}{16}$	1	$1\frac{5}{16}$	$1\frac{5}{8}$	$1\frac{5}{8}$	$2\frac{9}{16}$	$4\frac{1}{4}$
52	$\frac{5}{16}$	$\frac{17}{32}$	$\frac{27}{32}$	$\frac{27}{32}$	$1\frac{1}{16}$	$1\frac{5}{16}$	$1\frac{11}{16}$	$2\frac{1}{8}$	$2\frac{5}{8}$	$2\frac{5}{8}$	$4\frac{3}{16}$	$6\frac{15}{16}$
85	$\frac{1}{2}$	$\frac{27}{32}$	$1\frac{3}{8}$	$1\frac{3}{8}$	$1\frac{11}{16}$	$2\frac{1}{8}$	$2\frac{3}{4}$	$3\frac{1}{2}$	$4\frac{5}{16}$	$4\frac{5}{16}$	$6\frac{13}{16}$	$11\frac{1}{4}$
126	$\frac{3}{4}$	$1\frac{1}{4}$	2	2	$2\frac{1}{2}$	$3\frac{1}{8}$	$4\frac{1}{16}$	$5\frac{3}{16}$	$6\frac{7}{16}$	$6\frac{7}{16}$	$10\frac{1}{8}$	$16\frac{3}{4}$
203	$1\frac{3}{16}$	2	$3\frac{1}{4}$	$3\frac{1}{4}$	$4\frac{1}{16}$	$5\frac{1}{16}$	$6\frac{1}{2}$	$8\frac{3}{8}$	$10\frac{3}{8}$	$10\frac{3}{8}$	$16\frac{1}{4}$	27
330	2	$3\frac{5}{16}$	$5\frac{1}{4}$	$5\frac{1}{4}$	$6\frac{5}{8}$	$8\frac{1}{4}$	$10\frac{1}{2}$	$13\frac{1}{2}$	$16\frac{7}{8}$	$16\frac{7}{8}$	$26\frac{1}{2}$	44

Adjustable Dials, graduated to read to thousandths of an inch, indicate the transverse and vertical movements of table.

The Frame is hollow and fitted as a closet to hold the small parts that accompany the machine. On the left side of the frame there is a pan for holding small tools, etc., and on the front of this there is a rack for wrenches.

The Vise, with Rack Feed Machine, is flanged and has jaws $6\frac{1}{8}''$ wide, $1\frac{7}{16}''$ deep and opens $3\frac{5}{8}''$.

The Vise, with Screw Feed Machine, swivels and has a graduated base. The jaws are $5\frac{1}{8}''$ wide, $1\frac{1}{8}''$ deep, and will open $2\frac{3}{4}''$.

The Counter-Shaft has tight and loose pulleys, $14''$ in diameter, for $3\frac{1}{2}''$ belts, and should run about 180 revolutions per minute.

Weight of machine ready for shipment, about 2225 lbs.

Net Weight, about 1750 lbs.

Floor Space, Rack Feed Machine, $68'' \times 45''$; Screw Feed Machine, $79'' \times 45''$.

Dimensions of box in which machine is shipped, $50'' \times 34'' \times 62''$.

Each machine is furnished with vise, oil can, collet, wrenches, Treatise on Milling Machines, and everything shown in cut, together with overhead works.

MILLING TABLE WAYS IN BED.

No. 3

34 in. x 7 in. x 19 3-4 in.

PLAIN MILLING MACHINE.

The table has an automatic longitudinal feed of 34″, a transverse movement of 7″, and can be lowered 19¾″ from centre of spindle.

No. 3

34 in. x 7 in. x 19 3-4 in.

PLAIN MILLING MACHINE.

When desired, we furnish this machine fitted with an oil pump. When so fitted the oil channels are larger and a tank is provided.

The Spindle has a hole ¾" in diameter its entire length, and runs in bronze boxes provided with means of compensation for wear. For method of adjustment see page 17. The front end is threaded, and has a No. 11 taper hole.

The Cone has 3 steps, the largest 11" diameter, for 3½" belt, and is back geared, giving, with two speeds of counter-shaft, 12 changes of speed. The gears are covered.

The Overhanging Arm has a hole for a centre, or for a bearing for outer end of arbor, etc. It can be easily reversed to receive an attachment, turned out of the way, or removed. The distance from the centre of the spindle to the arm is 6⅜"; greatest distance from end of spindle to centre in arm, 21". An arm support is furnished and with this in position milling can be done to 15½" from face of column.

The Table, including oil pans and channels, is 50" long, 12" wide, 13¼" when fitted with oil pump, has a working surface 42" x 12", 3 T slots, ⅝" wide, a transverse movement of 7", 6" when fitted with oil pump, and can be lowered 19¾" from centre of spindle.

The Feed of table, 34", is automatic in either direction, and there are 16 changes of feed, obtained by transposing the feed pulleys, page 66, held in place by knurled nuts, varying from .004" to .332" to one revolution of spindle.

FEED TABLE FOR NO. 3 PLAIN MILLING MACHINE.

SPIN.

FEED SHAFT

STUD

FEED PER REV. OF SPINDLE	.004	.006	.009	.013	.018	.019	.027	.030	.040	.045	.063	.068	.095	.142	.212	.332

SPINDLE SPEEDS — TRAVEL OF TABLE IN INCHES PER MINUTE.

SPINDLE SPEEDS	.004	.006	.009	.013	.018	.019	.027	.030	.040	.045	.063	.068	.095	.142	.212	.332
23	$\frac{1}{32}$	$\frac{1}{8}$	$\frac{7}{32}$	$\frac{5}{16}$	$\frac{13}{32}$	$\frac{7}{16}$	$\frac{5}{8}$	$\frac{11}{16}$	$\frac{29}{32}$	$1\frac{1}{16}$	$1\frac{7}{16}$	$1\frac{9}{16}$	$2\frac{3}{16}$	$3\frac{1}{4}$	$4\frac{5}{8}$	$7\frac{5}{8}$
30	$\frac{1}{8}$	$\frac{3}{16}$	$\frac{7}{32}$	$\frac{3}{8}$	$\frac{17}{32}$	$\frac{9}{16}$	$\frac{13}{16}$	$\frac{29}{32}$	$1\frac{3}{16}$	$1\frac{1}{8}$	$1\frac{5}{8}$	$2\frac{1}{16}$	$2\frac{7}{8}$	$4\frac{1}{4}$	$6\frac{3}{8}$	10
34	$\frac{1}{8}$	$\frac{7}{32}$	$\frac{5}{16}$	$\frac{7}{16}$	$\frac{5}{8}$	$\frac{21}{32}$	$\frac{29}{32}$	1	$1\frac{3}{8}$	$1\frac{1}{2}$	$2\frac{1}{4}$	$2\frac{5}{16}$	$3\frac{1}{4}$	$4\frac{13}{16}$	$7\frac{3}{16}$	$11\frac{1}{4}$
44	$\frac{3}{16}$	$\frac{1}{4}$	$\frac{13}{32}$	$\frac{9}{16}$	$\frac{25}{32}$	$\frac{27}{32}$	$1\frac{3}{16}$	$1\frac{5}{16}$	$1\frac{3}{4}$	2	$2\frac{3}{4}$	3	$4\frac{3}{16}$	$6\frac{1}{4}$	$9\frac{9}{16}$	$14\frac{5}{8}$
52	$\frac{7}{32}$	$\frac{5}{16}$	$\frac{15}{32}$	$\frac{11}{16}$	$\frac{15}{16}$	1	$1\frac{3}{8}$	$1\frac{9}{16}$	$2\frac{1}{16}$	$2\frac{5}{16}$	$3\frac{1}{4}$	$3\frac{9}{16}$	$4\frac{15}{16}$	$7\frac{3}{8}$	11	$17\frac{1}{4}$
68	$\frac{9}{32}$	$\frac{13}{32}$	$\frac{5}{8}$	$\frac{7}{8}$	$1\frac{1}{4}$	$1\frac{5}{16}$	$1\frac{13}{16}$	$2\frac{1}{16}$	$2\frac{3}{4}$	$3\frac{1}{16}$	$4\frac{5}{16}$	$4\frac{5}{8}$	$6\frac{7}{16}$	$9\frac{5}{8}$	$14\frac{3}{8}$	$22\frac{1}{2}$
127	$\frac{1}{2}$	$\frac{3}{4}$	$1\frac{1}{8}$	$1\frac{5}{8}$	$2\frac{5}{16}$	$2\frac{7}{16}$	$3\frac{7}{16}$	$3\frac{13}{16}$	$5\frac{1}{16}$	$5\frac{11}{16}$	8	$8\frac{5}{8}$	$12\frac{1}{8}$	18	27	42
164	$\frac{21}{32}$	$\frac{31}{32}$	$1\frac{1}{2}$	$2\frac{1}{8}$	$2\frac{15}{16}$	$3\frac{1}{8}$	$4\frac{7}{16}$	$4\frac{15}{16}$	$6\frac{9}{16}$	$7\frac{3}{8}$	$10\frac{3}{8}$	$11\frac{1}{8}$	$15\frac{5}{8}$	$23\frac{1}{4}$	$34\frac{3}{4}$	$54\frac{1}{4}$
189	$\frac{3}{4}$	$1\frac{1}{8}$	$1\frac{11}{16}$	$2\frac{7}{16}$	$3\frac{3}{8}$	$3\frac{9}{16}$	$5\frac{3}{16}$	$5\frac{11}{16}$	$7\frac{9}{16}$	$8\frac{1}{2}$	$11\frac{1}{8}$	$12\frac{7}{8}$	18	$26\frac{3}{4}$	40	$62\frac{1}{2}$
244	$\frac{31}{32}$	$1\frac{1}{2}$	$2\frac{3}{16}$	$3\frac{3}{16}$	$4\frac{3}{8}$	$4\frac{5}{8}$	$6\frac{9}{16}$	$7\frac{5}{16}$	$9\frac{3}{4}$	11	$15\frac{3}{8}$	$16\frac{5}{8}$	$23\frac{1}{4}$	$34\frac{3}{4}$	$51\frac{1}{2}$	81
288	$1\frac{1}{8}$	$1\frac{3}{4}$	$2\frac{9}{16}$	$3\frac{3}{4}$	$5\frac{3}{16}$	$5\frac{1}{2}$	$7\frac{3}{4}$	$8\frac{5}{8}$	$11\frac{1}{2}$	13	$18\frac{1}{8}$	$19\frac{5}{8}$	$27\frac{1}{4}$	41	61	$95\frac{1}{2}$
371	$1\frac{1}{2}$	$2\frac{1}{4}$	$3\frac{5}{16}$	$4\frac{13}{16}$	$6\frac{11}{16}$	$7\frac{1}{16}$	10	$11\frac{1}{8}$	$14\frac{7}{8}$	$16\frac{3}{4}$	$23\frac{1}{4}$	$25\frac{1}{4}$	$35\frac{1}{4}$	$52\frac{1}{2}$	$78\frac{1}{2}$	123

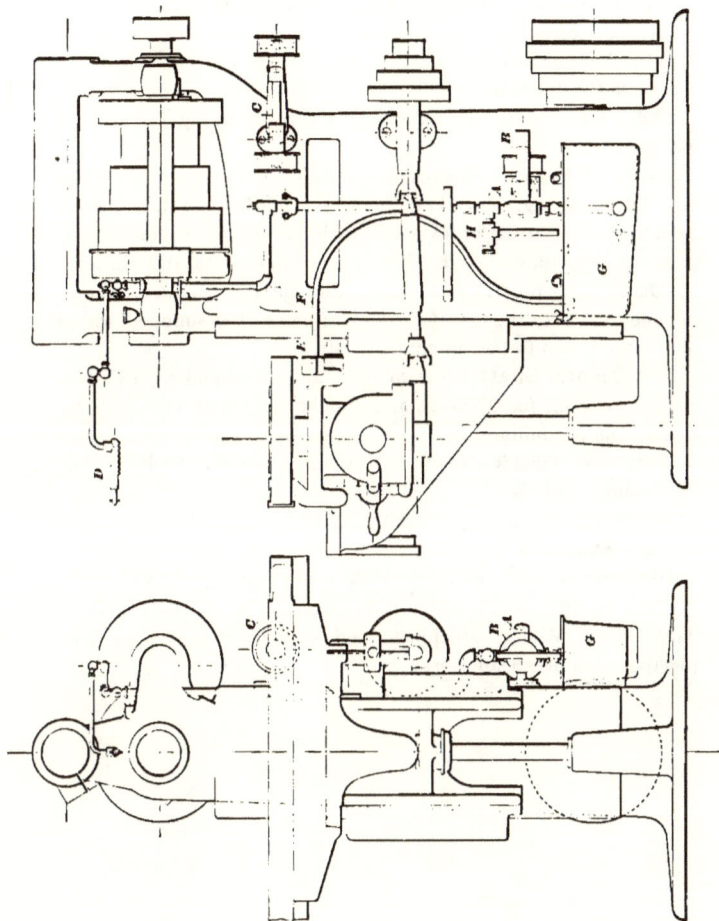

FIG. 18.

Adjustable Dials, graduated to read to thousandths of an inch, indicate the transverse and vertical movements of table.

The Frame is hollow and fitted as a closet to hold the small parts that accompany the machine. On the left side there is a pan for holding small tools, etc., and on the front of this there is a rack for wrenches.

The Vise is flanged and has jaws $6\frac{1}{8}''$ wide, $1\frac{7}{16}''$ deep, and will open $3\frac{3}{4}''$.

The Oil Pump and Fittings, and method of attaching the same are shown in Fig. 18. The pump A is driven by the pulley B from the pulley on the bracket C, which is in turn driven from a pulley on the counter-shaft. The oil is pumped to and through the distributing pipe D and returns through the conductor E and flexible tube F to the tank G. A relief valve H is supplied, which allows the pump to be run continuously.

The Counter-Shaft has two tight and loose pulleys, $14''$ and $18''$ in diameter, for $4''$ belts, and should run about 200 and 155 revolutions per minute.

Weight of machine ready for shipment, about 3275 lbs; with Oil Pump, 3380 lbs.

Net Weight, about 2710; with Oil Pump, 2800 lbs.

Floor Space, $84'' \times 51''$.

Dimensions of box in which machine is shipped, $56'' \times 32'' \times 66''$.

Each machine is furnished with vise, oil can, collet, wrenches, Treatise on Milling Machines, and everything shown in cut, together with overhead works.

MILLING SLOTS WITH SIDE MILLS.

No. 4

42 in. x 8 1-4 in. x 22 in.

PLAIN MILLING MACHINE.

(No. 6, Prior to 1893.)

The table has an automatic longitudinal feed of 42″, a transverse movement 8¼″, and can be lowered 22″ from centre of spindle.

No. 4

42 in. x 8 1-4 in. x 22 in.

PLAIN MILLING MACHINE.

(No 6, Prior to 1893.)

When desired, we furnish this machine fitted with an oil pump. When so fitted the oil channels are larger and a tank is provided.

The Spindle has a hole $\frac{3}{4}''$ in diameter its entire length, and runs in bronze boxes provided with means of compensation for wear. For method of adjusting see page 17. The front end is threaded, and has a No. 11 taper hole.

The Cone has 3 steps, the largest 13'' diameter, for $3\frac{1}{2}''$ belt, and is back geared, giving, with two speeds of counter-shaft, 12 changes of speed. The gears are inside the column.

The Overhanging Arm has a hole for a centre, or for a bearing for outer end of arbor, etc. It can be easily reversed to receive an attachment, turned out of the way, or removed. The distance from the centre of the spindle to the arm is $7\frac{1}{4}''$; greatest distance from end of spindle to centre in arm, $26\frac{1}{2}''$. An arm support is furnished and with this in position milling can be done to $18\frac{1}{2}''$ from face of column.

The Table, including oil pans and channels, is 60'' long, 14'' wide, $15\frac{7}{8}''$ when fitted with oil pump, has a working surface 48'' x 14'', 3 T slots, $\frac{3}{4}''$ wide, a transverse movement of $8\frac{1}{4}''$, $7\frac{1}{8}''$ when fitted with oil pump, and can be lowered 22'' from centre of spindle.

The Feed of table, 42'', is automatic in either direction, and there are 12 changes of feed, obtained by transposing the feed pulleys, page 73, held in place by knurled nuts, varying from .005'' to .236'' to one revolution of spindle.

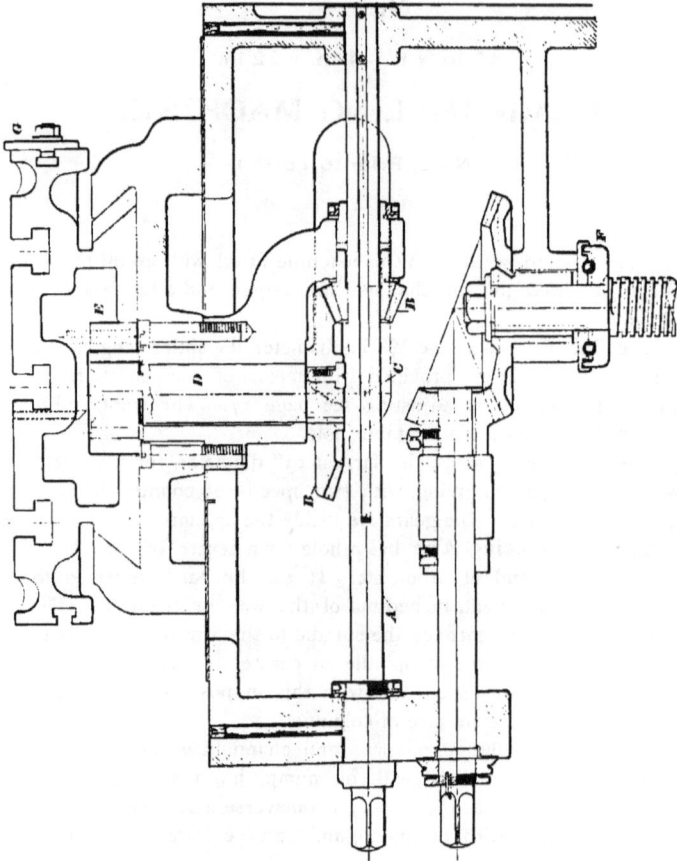

FIG. 19.

FEED TABLE FOR NO. 4 PLAIN MILLING MACHINE.

SPINDLE SPEEDS	FEED PER. REV. OF SPINDLE											
	.005	.007	.011	.017	.027	.028	.040	.041	.062	.099	.159	.236
	TRAVEL OF TABLE IN INCHES PER MINUTE											
14	$\frac{1}{16}$	$\frac{3}{32}$	$\frac{5}{32}$	$\frac{1}{4}$	$\frac{3}{8}$	$\frac{13}{32}$	$\frac{9}{16}$	$\frac{9}{16}$	$\frac{7}{8}$	$1\frac{3}{8}$	$2\frac{1}{4}$	$3\frac{5}{16}$
18	$\frac{3}{32}$	$\frac{1}{8}$	$\frac{3}{16}$	$\frac{5}{16}$	$\frac{1}{2}$	$\frac{1}{2}$	$\frac{23}{32}$	$\frac{3}{4}$	$1\frac{1}{8}$	$1\frac{3}{4}$	$2\frac{5}{8}$	$4\frac{1}{4}$
23	$\frac{1}{8}$	$\frac{5}{32}$	$\frac{1}{4}$	$\frac{13}{32}$	$\frac{5}{8}$	$\frac{21}{32}$	$\frac{29}{32}$	$\frac{15}{16}$	$1\frac{7}{16}$	$2\frac{1}{4}$	$3\frac{11}{16}$	$5\frac{7}{16}$
30	$\frac{5}{32}$	$\frac{7}{32}$	$\frac{11}{32}$	$\frac{1}{2}$	$\frac{13}{16}$	$\frac{27}{32}$	$1\frac{3}{16}$	$1\frac{1}{4}$	$1\frac{7}{8}$	3	$4\frac{3}{4}$	$7\frac{1}{16}$
39	$\frac{3}{16}$	$\frac{9}{32}$	$\frac{7}{16}$	$\frac{21}{32}$	$1\frac{1}{16}$	$1\frac{1}{16}$	$1\frac{9}{16}$	$1\frac{5}{8}$	$2\frac{7}{16}$	$3\frac{5}{8}$	$6\frac{3}{16}$	$9\frac{1}{4}$
50	$\frac{1}{4}$	$\frac{11}{32}$	$\frac{9}{16}$	$\frac{7}{8}$	$1\frac{3}{8}$	$1\frac{3}{8}$	2	$2\frac{1}{16}$	$3\frac{5}{8}$	$4\frac{15}{16}$	$7\frac{15}{16}$	$11\frac{3}{4}$
112	$\frac{9}{16}$	$\frac{25}{32}$	$1\frac{1}{4}$	$1\frac{7}{8}$	3	$3\frac{1}{8}$	$4\frac{1}{2}$	$4\frac{9}{16}$	$6\frac{15}{16}$	$11\frac{1}{8}$	$17\frac{3}{4}$	$26\frac{1}{2}$
144	$\frac{23}{32}$	1	$1\frac{9}{16}$	$2\frac{7}{16}$	$3\frac{7}{8}$	$4\frac{1}{16}$	$5\frac{3}{4}$	$5\frac{7}{8}$	$8\frac{15}{16}$	$14\frac{1}{4}$	23	34
186	$\frac{15}{16}$	$1\frac{5}{16}$	$2\frac{1}{16}$	$3\frac{3}{16}$	5	$5\frac{3}{16}$	$7\frac{1}{2}$	$7\frac{5}{8}$	$11\frac{1}{4}$	$18\frac{3}{8}$	$29\frac{1}{4}$	44
240	$1\frac{3}{16}$	$1\frac{11}{16}$	$2\frac{5}{8}$	$4\frac{1}{16}$	$6\frac{1}{2}$	$6\frac{3}{4}$	$9\frac{5}{8}$	$9\frac{7}{8}$	$14\frac{7}{8}$	$23\frac{3}{4}$	$38\frac{1}{4}$	$56\frac{1}{2}$
310	$1\frac{9}{16}$	$2\frac{3}{16}$	$3\frac{7}{16}$	$5\frac{1}{4}$	$8\frac{3}{8}$	$8\frac{5}{8}$	$12\frac{3}{8}$	$12\frac{3}{4}$	$19\frac{1}{4}$	$30\frac{1}{2}$	$49\frac{1}{2}$	73
400	2	$2\frac{11}{16}$	$4\frac{3}{8}$	$6\frac{13}{16}$	$10\frac{3}{4}$	$11\frac{1}{4}$	16	$16\frac{3}{8}$	$24\frac{3}{4}$	$39\frac{1}{2}$	$63\frac{1}{2}$	$94\frac{1}{2}$

Figure 19 shows a cross section of the table and knee of a Plain Milling Machine. The longitudinal hand feed is through the shaft A, bevel gears B B, and the shaft D, to which the pinion is attached, engaging in the rack E. The power feed is driven from the side through worm and worm gears to a bevel gear, the centre of which is at C, engaging in the bevel gear B, and driven as by hand. A T slot on the back of the table is provided for bolts to hold stops, against which work can be placed when it is desired to locate it from the back of the table, G.

Adjustable Dials, graduated to read to thousandths of an inch, indicate the transverse and vertical movements of table.

Oil Pump. For sketch and description of method of applying oil pump see pages 67 and 68.

The Frame is hollow and fitted as a closet to hold the small parts that accompany the machine. On the left side there is a pan for holding small tools, etc., and on the front of this there is a rack for wrenches.

The Vise is flanged and has jaws $7\frac{1}{8}''$ wide, $1\frac{7}{8}''$ deep, and will open $4\frac{1}{2}''$.

The Counter-Shaft has two tight and loose pulleys, $14''$ and $18''$ in diameter, for $4''$ belts, and should run about 200 and 155 revolutions per minute.

Weight of machine ready for shipment, about 4100 lbs; with Oil Pump, 4470 lbs.

Net Weight, about 3590 lbs; with Oil Pump, 3740 lbs.

Floor Space, $102'' \times 59''$.

Dimensions of box in which machine is shipped, $62'' \times 39'' \times 67''$.

Each machine is furnished with vise, oil can, collet, wrenches, Treatise on Milling Machines, and everything shown in cut, together with overhead works.

NO. 5 PLAIN MILLING MACHINE IN OPERATION.

No 5
48 in. x 9 3-4 in. x 19 1-2 in.
PLAIN MILLING MACHINE.
(No. 8, Prior to 1893.)

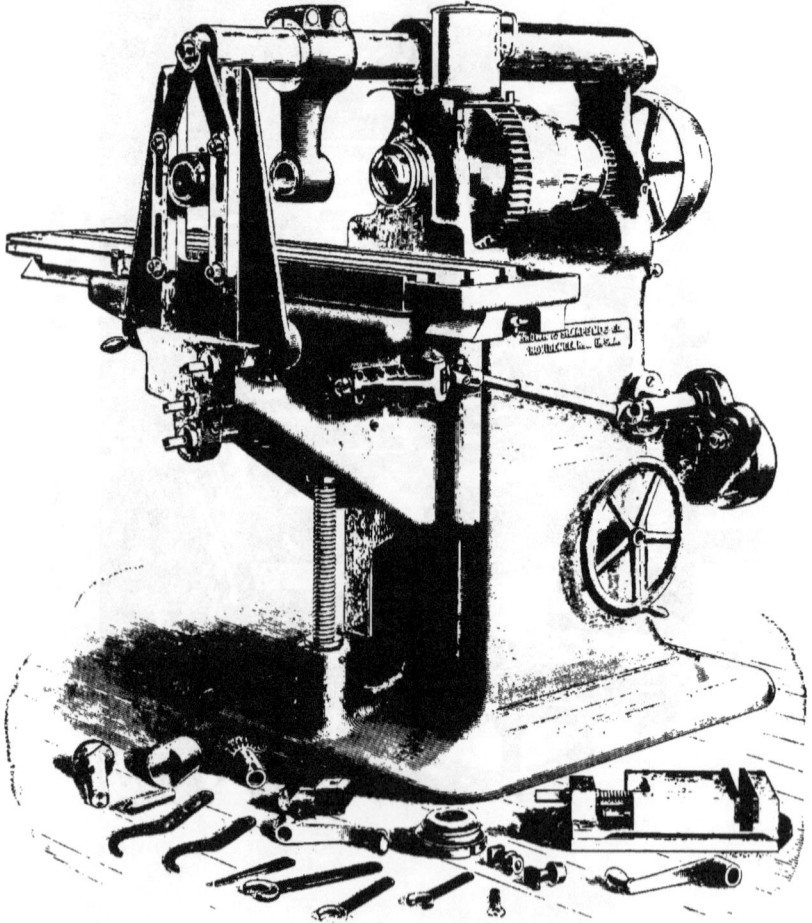

The table has an automatic longitudinal feed of 48″, a transverse movement of 9¾″, and can be lowered 19½″ from centre of spindle.

No 5

48 in. x 9 3-4 in. x 19 1-2 in.

PLAIN MILLING MACHINE.

(No. 8, Prior to 1893.)

Patented Oct. 18, 1892.

When desired, we furnish this machine fitted with an oil pump.
When so fitted the oil channels are larger and a tank is provided.

The Spindle has a hole ¾" in diameter its entire length, and
runs in bronze boxes provided with means of compensation for
wear. The front end has a No. 12 taper hole.

The Cone has 3 steps, the largest 13¾" diameter, for 4½"
belt, and is back geared, giving, with two speeds of counter-shaft,
12 changes of speed. The gears are inside the column.

The Overhanging Arm has a hole for a centre, or for a bearing for outer end of arbor, etc. It can be easily pushed back
from the table. The distance from the centre of the spindle to
the arm is 8"; greatest distance from end of spindle to centre in
arm, 24". An arm support is furnished, and with this in position,
milling can be done to 26" from face of column.

The Table, including oil pans and channels, is 66¼" long, 16"
wide, 18" when fitted with oil pump, and has a working surface
54" x 16", 3 T slots, ¾" wide, a transverse movement of 9¾",
8¼" when fitted with oil pump, and can be lowered 19½" from
centre of spindle.

The Feed of table, 48" is automatic in either direction, and
there are 8 changes of feed, obtained by transposing the feed
pulleys, held in place by knurled nuts, varying from .023" to .25"
to one revolution of spindle.

Adjustable Dials, graduated to read to thousandths of an inch, indicate the transverse and vertical movements of table.

Oil Pump. For sketch and description of method of applying see pages 67 and 68.

The Frame is hollow and fitted as a closet to hold the small parts that accompany the machine. On the left side there is a pan for holding small tools, etc., and on the front of this there is a rack for wrenches.

The Vise is flanged and has jaws $7\frac{1}{8}''$ wide, $1\frac{7}{8}''$ deep, and will open $4\frac{1}{2}''$.

The Counter-Shaft has two tight and loose pulleys, $16''$ and $20''$ in diameter, for $5''$ belts, and should run about 140 and 112 revolutions per minute.

Weight ready for shipment, 6160 lbs.; with Oil Pump, 6400 lbs.

Net Weight, about 5150 lbs.; with Oil Pump, 5250 lbs.

Floor Space, $115'' \times 69''$.

Dimensions of boxes in which machine is shipped, $67'' \times 42'' \times 67''$ and $66'' \times 21'' \times 21''$.

Each machine is provided with vise, oil can, collet, wrenches, Treatise on Milling Machines, and everything shown in cut, together with overhead works.

MILLING STEEL MOTOR PINIONS.

No. 12

17 1-2 in. x 5-8 in. x 8 in.

PLAIN MILLING MACHINE.

(No. 2, Prior to 1893.)

The table has an automatic longitudinal feed of 17½", the spindle has a transverse adjustment of 5⅝", and the greatest distance from centre of spindle to top of table is 8".

No. 12

17 1-2 in. x 5-8 in. x 7 1-2 in.

PLAIN MILLING MACHINE.

(No. 2, Prior to 1893.)

———————

This machine by reason of its rigidity and the ease with which the table can be moved by hand makes it well adapted for the rapid and economical production of many parts of machine tools, bicycles, electrical instruments, guns, etc.

The Spindle runs in bronze boxes provided with means of compensation for wear. It is driven from cone by gear and pinion, has a vertical adjustment by means of nuts placed on a vertical screw, and a transverse adjustment of $\frac{5}{8}''$. The front end has a No. 10 taper hole.

The Cone has 3 steps, the largest $12\frac{1}{8}''$ in diameter, for $2\frac{1}{2}''$ belt.

The Overhanging Arm has an adjustable centre support and brace. Distance from centre of spindle to arm, $3\frac{3}{16}''$; greatest distance from end of spindle to centre in arm, with arm brace in position, $7\frac{3}{4}''$, without arm brace, $9''$.

The Table, including oil pans and channels, is $28''$ long and $10''$ wide, has a working surface $19'' \times 6''$, and a T slot $\frac{5}{8}''$ wide. Greatest distance from centre of spindle to top of table, $7\frac{1}{2}''$, least, $2\frac{1}{2}''$.

The Feed of table, of $17\frac{1}{2}''$, is automatic and can be automatically released at any point. There are three changes of feed, varying from $.014''$ to $.05''$ to one revolution of spindle.

In addition to the oil pans and channels surrounding the table, an oil tank is attached to each machine providing for the use of a pump.

The Vise has jaws $6\frac{1}{4}''$ wide, $1\frac{7}{16}''$ deep and will open $3\frac{5}{8}''$.

The Counter-Shaft has tight and loose pulleys $10''$ in diameter for $3''$ belts, and should run about 280 revolutions per minute.

Weight of machine ready for shipment; about 1950 lbs.

Net Weight, about 1600 lbs.

Floor Space, $46'' \times 45''$.

Dimensions of box in which machine is shipped, $43'' \times 34'' \times 57''$.

Each machine is provided with vise, oil can, wrenches, Treatise on Milling Machines, and everything shown in cut, together with overhead works.

MILLING TABLE WAYS.

No. 12
PLAIN MILLING MACHINE.

Arranged for Sprocket Wheel Cutting.

MACHINES FOR BICYCLE WORK.

Development of the bicycle manufacturing business has called for a great variety of machines specially adapted for bicycle work.

On page 83 is shown our No. 12 Plain Milling Machine, described in detail on pages 81 and 82, fitted with single dial index centres. By this method gangs of sprocket wheels can be placed upon arbors, and by using a cutter specially made for the purpose two teeth can be cut at one time.

The Index Plate is provided with hardened steel taper bushings, and is covered, thus protecting the holes from dirt. A hardened steel taper pin is forced into the bushing by a spring, and can be released by a lever, when the work can be rotated by a hand wheel, thus making the indexing very rapid. While the plates can be used, usually, for other than the number of teeth for which they are made, it is desirable to have them contain holes for the actual number of teeth to be cut, as mistakes can then be avoided.

The Foot-Stock is provided with a bearing instead of a centre, which gives the arbor a firm support.

Gangs of sprockets from 6″ to 7″ long can be milled.

If single cutters are used, cutting one tooth at a time, a set of four cutters will cut 7, 8, 17, 18, 19 and 20-tooth sprockets.

We also show on page 85 the same machine fitted for cutting chain centres. By means of a special holding device, a number of these centres can be cut at one time, the number being limited by the size of the chain and the distance between the face of the spindle and the centre in the arm.

Other operations can be profitably done on some of the other machines described in the preceding pages.

By consultation with those interested in bicycle manufacture, many methods may be designed by which the work may be greatly facilitated.

NO. 12 PLAIN MILLING MACHINE ARRANGED FOR CUTTING
BICYCLE CHAIN CENTRES.

No. 13

15 in. x 3 in. x 6 3-8 in.

PLAIN MILLING MACHINE.

(No. 3, Prior to 1893.)

The table has an automatic longitudinal feed of 15″, a transverse movement of 3″, and the greatest distance from centre of spindle to top of table is 6⅜″.

No. 13

15 in. x 3 in. x 6 3-8 in.

PLAIN MILLING MACHINE.

(No. 3, Prior to 1893.)

This machine combines power and rigidity with ease of handling, making it especially valuable in milling the heavier cuts on bicycle and sewing machine parts and other work of this class.

The Spindle runs in boxes provided with means of compensation for wear. It is driven from cone by gear and pinion and has a vertical adjustment by means of nuts placed on a vertical screw. The front end has a No. 10 taper hole.

The Cone has 3 steps, the largest 13″ in diameter, for 3″ belt.

The Overhanging Arm has an adjustable centre support and an arm brace. Distance from centre of spindle to arm, 2⅞″; greatest distance from end of spindle to centre in arm, 11″.

The Table, including oil pans and channels, is 31½″ long and 10½″ wide, has a working surface 27″ x 8″, 2 T slots ¾″ wide and a transverse movement of 3″. Greatest distance from centre of spindle to top of table, 6⅜″; least, 3½″.

The Feed of table, of 15″, is automatic and can be automatically released at any point. It is a screw feed and can be quickly returned by hand. There are three changes of feed, varying from .015″ to .04″ to one revolution of spindle.

In addition to the oil pans and channels surrounding the table, an oil tank is attached to each machine providing for the use of a pump.

88   BROWN & SHARPE MFG. CO.

The Vise has jaws $6\frac{1}{4}''$ wide, $1\frac{7}{16}''$ deep, and will open $3\frac{3}{4}''$.

The Counter-Shaft has tight and loose pulleys $10''$ in diameter for $3\frac{1}{4}''$ belt, and should run about 375 revolutions per minute.

Weight of machine ready for shipment, about 2650 lbs.

Net Weight, about 2240 lbs.

Floor Space, $47'' \times 50''$.

Dimensions of box in which machine is shipped, $50'' \times 38'' \times 59''$.

Each machine is furnished with vise, oil can, wrenches, Treatise on Milling Machines, and everything shown in cut, together with overhead works.

PLAIN MILLING.

MILLING CARRIAGE WAY.

MILLING HANGER BOXES.

No. 23
49 in. x 9 in. x 19 in.
PLAIN MILLING MACHINE.
(No. 5, Prior to 1893.)

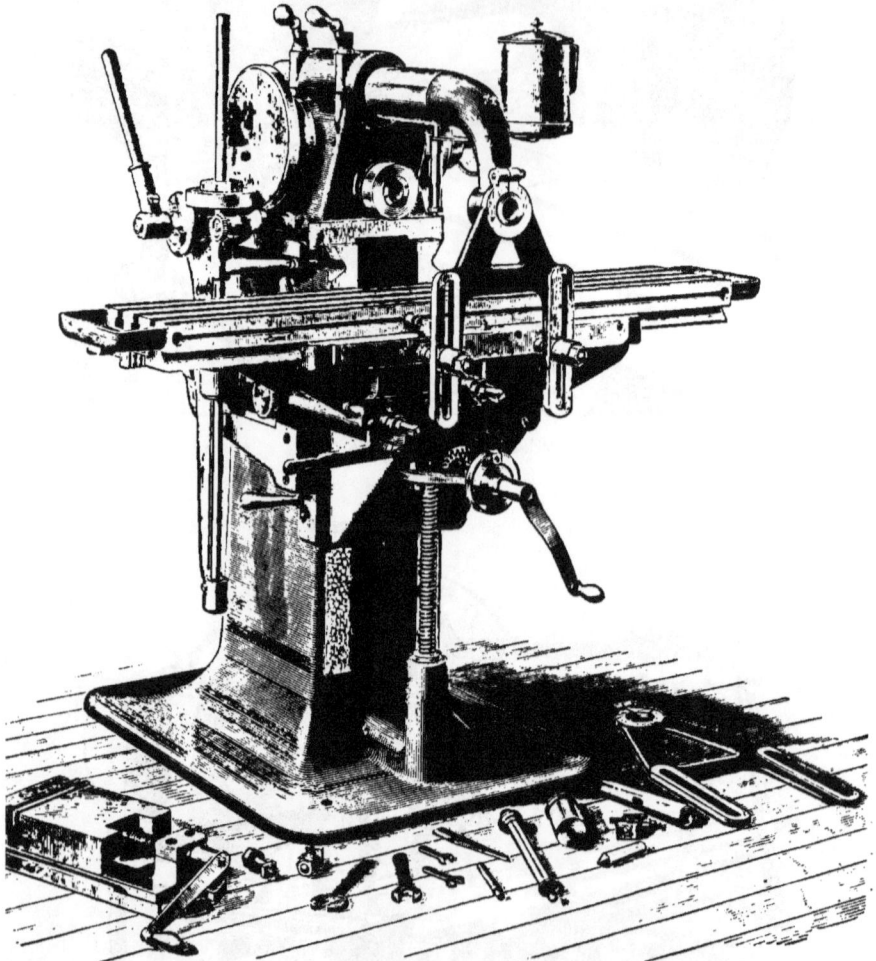

The table has an automatic feed of 49″, the head has a transverse movement of 9″, and the table can be lowered 19″ from centre of spindle.

No. 23

49 in. x 9 in. x 19 in.

PLAIN MILLING MACHINE.

(No. 5, Prior to 1893.)

This machine is especially adapted for work requiring large table capacity and long cuts.

The Spindle is hollow, runs in bronze boxes provided with means of compensation for wear, and is driven by a worm and worm gear. The worm is of steel, hardened, and the worm gear of bronze. The worm gear runs in oil. The front end of spindle is threaded and has a No. 11 taper hole.

The Cone has 3 steps, the largest $10\frac{3}{8}''$ in diameter, for $2\frac{1}{2}''$ belt, and with two speeds of counter gives 6 changes of speed.

The Head has a transverse movement of $9''$.

The Overhanging Arm can be removed or turned out of the way. Distance from centre of spindle to arm, $5\frac{1}{2}''$; greatest distance from end of spindle to centre in arm, $15''$. A long and a short arm support are furnished and with one of these in position milling can be done to $15''$ from face of column.

The Table, including oil pans and channels, is $59''$ long and $10''$ wide, has a working surface $50\frac{3}{4}'' \times 10''$, 3 T slots, $\frac{3}{4}''$ wide and can be lowered $19''$ from centre of spindle.

The Feed of table, of $49''$, is automatic in either direction and can be automatically released at any point. It is driven by a friction disk with connections that automatically adjust themselves to any position of head and knee, and can be quickly changed from 0 to $.10''$ to one revolution of spindle.

Dials graduated to read to 16ths, 32nds, etc., and to thousandths of an inch indicate the transverse movement of head and vertical movement of table.

The Vise is flanged and has jaws 7¼" wide, 1⅞" deep, and will open 4½".

The Counter-Shaft has tight and loose pulleys 10" and 14" in diameter for 3" and 3½" belts, and should run about 450 and 300 revolutions per minute.

Weight of machine ready for shipment, about 3100 lbs.

Net Weight, about 2600 lbs.

Floor Space, 45" x 110".

Dimensions of box in which machine is shipped, 50" x 37" x 65".

Each machine is furnished with vise, oil can, collet, wrenches, Treatise on Milling Machines, and everything shown in cut, together with overhead works.

MILLING T SLOTS IN TABLE.

No. 24

60 in. x 12 in. x 17 5-8 in. and
72 in. x 12 in. x 17 5-8 in.

PLAIN MILLING MACHINES.

(No. 7, Prior to 1893.)

The No. 24 Plain Milling Machine is similar in general design and construction to the No. 23 Plain Milling Machine. All the parts are larger and heavier, however, and thus greater capacity and efficiency are obtained.

The Spindle is hollow, runs in bronze boxes provided with means of compensation for wear, and is driven by a worm and worm gear. The worm is of steel, hardened, and the worm gear of bronze. The worm gear runs in oil. The front end of spindle is threaded and has a No. 12 taper hole.

The Cone has 3 steps, the largest 12″ in diameter, for 3″ belt, and with two speeds of counter, gives 6 changes of speed.

The Head has a transverse movement of 12″.

The Overhanging Arm can be removed or turned out of the way. Distance from centre of spindle to arm, $5\frac{9}{16}''$; greatest distance from end of spindle to centre in arm, 18″. A long and short arm support are furnished, and with one of these in position milling can be done to $17\frac{1}{2}''$ from face of column.

The Table, including oil pans and channels, is 69″ long and $14\frac{1}{2}''$ wide, has a working surface 60″ x $14\frac{1}{2}''$, 5 T slots $\frac{3}{4}''$ wide, and can be lowered $17\frac{5}{8}''$ below the centre of spindle.

The Feed of table, of 60″, is automatic in either direction. It is driven by a friction disk with connections that automatically adjust themselves to any position of head and knee, and can be quickly changed from o to .08″ to one revolution of spindle.

Dials graduated to read to 16ths, 32nds, etc., and to thousandths of an inch indicate the transverse movement of head and vertical movement of table.

The Vise is flanged and has jaws $7\frac{1}{4}''$ wide, $1\frac{7}{8}''$ deep, and will open $4\frac{1}{2}''$.

The Counter-Shaft has tight and loose pulleys $12''$ and $18''$ in diameter for $3\frac{1}{2}''$ and $4\frac{1}{2}''$ belts, and should run about 300 and 200 revolutions per minute.

Weight of machine ready for shipment, about 4325 lbs.

Net Weight, about 3700 lbs.

Floor Space, $52'' \times 129''$.

Dimensions of box in which machine is shipped, $63'' \times 39'' \times 67''$. This machine is also furnished with table $72''$ long.

Weight of machine ready for shipment, about 4700 lbs.

Net Weight, about 3950 lbs.

Floor Space, $52'' \times 153''$.

Dimensions of boxes in which machine with this table is shipped, $57'' \times 51'' \times 68''$ and $77'' \times 19'' \times 8''$.

Each machine is furnished with vise, oil can, collet, wrenches, Treatise on Milling Machines, and overhead works.

No. 2
41 in. x 12 in. x 15 in.
Vertical Spindle Milling Machine.

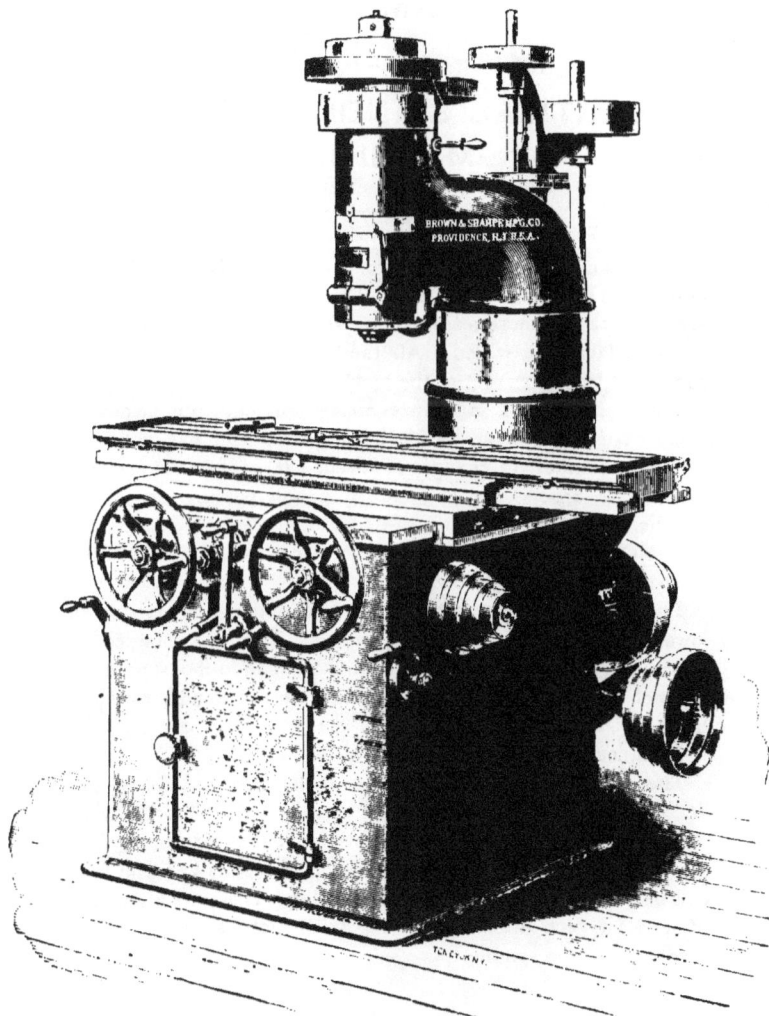

This machine has a longitudinal feed of 41″ and a transverse feed of 12″. Greatest distance from end of spindle to table, 15″.

No. 2

41 in. x 12 in. x 15 in.

VERTICAL SPINDLE MILLING MACHINE.

This machine, for many kinds of work, is preferable to a machine with a horizontal spindle. The operator can more easily see the work and more readily follow any irregularity in the outline of the surface to be milled. All the movements of the table are controlled from the front of the machine.

The Spindle runs in bronze boxes provided with means of compensation for wear. It is back geared and has, with two speeds of counter, 12 changes of speed. It has a No. 11 taper hole. The arbors can be held by a bolt passing through the spindle.

The Cone has 3 steps for 3" belt, and can be set parallel with, or at right angles to the bed.

The Vertical Adjustment of 13½" is obtained by raising or lowering the column ; a fine adjustment of $1\frac{5}{16}$" is obtained by means of a collar nut that is graduated to read to thousandths of an inch. The greatest distance, from end of spindle to top of table, 15", the least, 1½".

Distance from centre of spindle to column, 13½".

The Table, including oil pans and channels, is 53" long and 16" wide, has a working surface 41" x 13¾", 3 T slots ¾" wide.

The Feeds of table are automatic in either direction. The longitudinal feed is 41" and the transverse feed is 12", and both can be automatically released at any point. There are 8 changes of feed for each direction varying from .006" to .056" to one revolution of the spindle.

The **Counter-shaft** has friction pulleys 12" and 16" in diameter for 3½" and 4" belts, and should run about 220 and 165 revolutions per minute.

Weight of machine ready for shipment, about 5250 lbs.

Net Weight, about 4400 lbs.

Floor Space, 100"x 78".

Dimensions of box in which each machine is shipped, 69" x 45" x 83".

Each machine is furnished with collet, oil can and stand, wrenches, Treatise on Milling Machines, and overhead works.

CIRCULAR MILLING ATTACHMENT.

This Attachment, for the No, 2 Vertical Spindle Milling Machine, is of service in milling circles, segments of circles, circular slots, etc., on plain and irregularly shaped pieces. It is bolted to the table of the machine and when so placed can be adjusted to any desired position.

By the addition of this attachment, the machine is fully equipped to do all varieties of straight and circular milling within its capacity.

The Table is 20" in diameter and has 6 T slots ¾" wide.

The Feed of table is automatic and can be automatically released at any point. There are 8 changes of feed.

The Attachment is 4¼" high.

Weight ready for shipment, about 325 lbs.

Net Weight, about 250 lbs.

Each Attachment is furnished with bracket and pulleys for attaching it to the machine.

GEAR CUTTING ATTACHMENT.

For No. 1 Universal Milling Machine, Design Prior to 1895.

This attachment is for use in cutting gear wheels, and wheels larger and heavier than can be cut with the ordinary apparatus belonging to the No. 1 Universal Milling Machine, design prior to 1895. It swings 14″, and is furnished with a 20″ index plate containing 4294 holes. It will divide all numbers to 75, and all even numbers to 150. Arbors fitted to the No. 1 Universal Milling Machine can be used in this attachment. The screw with set nuts over the spindle is designed as a support for the wheel while being cut. Weight, boxed, 440 lbs.

The Head and Foot-Stock of this attachment can be fitted for use upon the Nos. 1, design of 1895, 2, 3 and 4, design of 1893, Universal, and the Nos. 1, 2, 3, 4, 5, 23 and 24 Plain Milling Machines.

VERTICAL SPINDLE MILLING ATTACHMENTS.

For Nos. 1, 2, 3 and 4 Universal, and Nos. 1, 2, and 3, Plain Milling Machines.

These attachments are used for a large range of light milling, and are of special advantage for key-seating, die-sinking, cutting T slots, spiral gears or worms, sawing stock when in lengths greater than can be placed at right angles to the table, etc.

The holder or frame is secured to the overhanging arm, and the horizontal shaft is inserted in the cone spindle of the machine. The vertical spindle is driven by the horizontal shaft through spiral gears.

The spindle can be set at any angle from a vertical to a horizontal position. The position is indicated on the base of spindle head which is graduated.

Machine on which at- } Nos. 1 & 2 Univ. Nos. 3 & 4 Univ.
tachment is used,   } Nos. 1 & 2 Plain. No. 3 Plain.

Taper Hole in Spindle,   7   9

VERTICAL SPINDLE MILLING ATTACHMENT ARRANGED TO SAW OFF
STOCK.

VERTICAL SPINDLE MILLING ATTACHMENT ARRANGED FOR CUTTING SPIRAL ROLLS.

VERTICAL SPINDLE MILLING
ATTACHMENTS

For Nos. 4 and 5 Plain and No. 4, Design 1893, Universal Milling Machines.

The holder or frame is secured to the frame of the machine, and the horizontal shaft is inserted in the spindle of the machine.

The vertical spindle is driven by the horizontal shaft through bevel gears.

The spindle can be·set at any angle from a vertical to a horizontal position. The position is indicated on the base of spindle head, which is graduated.

Machine on which Attachment is used.	No. 4 Plain and No. 4 Univ. Design 1893.	No. 5, Plain.
Taper Hole in Spindle,	9	11
Distance from Centre of spindle to Column,	9½ in.	11 in.

VERTICAL SPINDLE MILLING ATTACHMENT

For No. 24 Plain Milling Machines.

The holder or frame is secured to the overhanging arm, and the vertical spindle is driven by a worm and a worm wheel by a belt from cone of the machine.

On certain classes of work two cuts can be made simultaneously, one by a cutter on the arbor in the spindle of the machine, the other by a mill in the vertical spindle of the attachment.

HIGH SPEED MILLING ATTACHMENT AND DRIVING FIXTURE

For Nos. 1 and 2 Universal and 1 and 2 Plain Milling Machines.

The cut on the left represents the **High Speed Milling Attachment**, which is of service for light or finishing cuts with end mills, and consists of a small collet or spindle running in a hardened shell that fits the hole in the cone spindle of the machine. It can

DRIVING FIXTURE FOR HIGH SPEED MILLING ATTACHMENT.

Method of Using when a Quick Feed is required for the Table.

be driven directly from an extra pulley on the counter-shaft, and thus small mills may be run much faster than the usual speed of the cone spindle. When the attachment is thus driven, the cone pulley is stationary.

The cut on the right represents the **Driving Fixture,** which consists of a cast iron arm that can be readily placed in a milling machine, in the place of the overhanging arm, and can be applied without additional overhead works.

There is a shaft running in the upper end that has a pulley on each end of it. The pulley on the back is to be belted to the front step of the cone of the machine. The front pulley is to be belted to the fixture that is to be placed in the taper hole of the spindle.

The advantage of this method of driving is that the steps of the cone pulley are free to be run in the usual manner, and thus different speeds can be obtained for the end mill.

On the No. 1 Universal and No. 1 Plain Milling Machine there are 4 changes of speed, from 342 to 1537 revolutions per minute, and on the No. 2 Universal and No. 2 Plain Milling Machines there are 3 changes of speed, from 603 to 1584 revolutions per minute.

HAND MILLING ATTACHMENT

For No. 0 Plain Milling Machine.

The No. 0 Plain Milling Machine can be quickly changed by means of this attachment into a hand milling machine with or without an automatic longitudinal feed.

An apron, placed on the outside end of the knee, carries a lever attached to a segment of a gear which runs in a pinion placed over the end of the shaft that moves the table longitudinally, and this lever when moved turns the shaft as the crank would if it were in position.

The attachment, with a knee having a working surface of 6″ x 5¾″, is clamped on the table and on this the fixtures for holding the work can be fastened as on a hand milling machine. When brought to position the lever can be held by the catch in the holder, shown at the left of the cut, which can be released by a latch on the back of the lever, so that at the same time that the knee is returned to position the catch is released without an extra movement. While the lever is held down, the feed can be thrown in and milling done as on a plain milling machine.

The top of the knee at its lowest position is 6″ from the top of the table and can be raised 2″.

With this attachment in position the milling machine table has a transverse feed of 2¼″. The longitudinal feed of the table by means of the lever and gear segment is 4″, but with these removed the machine will feed 16″ automatically.

TAPER MILLING ATTACHMENT
For Nos. 1 and 2 Universal Milling Machines.

This attachment is designed to facilitate the milling of taper work. By reason of its easy and quick adjustment to the desired taper it is especially desirable when a large variety of such work is to be done.

It consists of a table that is suspended on a ring, which in turn is placed on an arbor to fit the taper hole in the spiral head. The head can be set to any desired angle to 10°, and the table will take the same position, keeping the centres always in line. When placed at the required angle it is held in position by a clamp screw that slides in a knee clamped to the table of the machine. By reason of the location of the clamps and the solidity of the table, work is held as firmly between the centres of this attachment as they are between centres fastened directly to the table.

The foot-stock of the attachment slides in a T slot 5⅛″ wide, and can be placed to take in work to 4¼″ in diameter and 17″ in length.

In ordering state whether it is to be used on the No. 1 or No. 2 Universal Milling Machine.

INDEX HEAD AND CENTRES WITH CENTRE REST.

PATENTED FEBRUARY 5, 1884.

These Centres are designed for use on Milling Machines, but the strength, stiffness and stability of the Index head in all positions enable it also to be advantageously used on Planers, Upright Drills and Slotting Machines.

The Centres swing 12½" in diameter.

The Head can be set at any angle from 10 degrees below the horizontal to 10 degrees beyond the perpendicular.

The Spindle is provided with a face plate and adjustable dog carrier. The front end has a No. 12 taper hole. The straight hole at end of taper is 1⅛" in diameter.

The Worm Wheel is 6" in diameter, and one revolution is made by 60 revolutions of index crank.

The Index Plates divide all numbers to 100, all even numbers to 134, and all divisible by 4 to 200.

The Table is provided with flanges, is 32" long, 8" wide, and has 3 T slots ¾" wide.

Combined Length of head and foot-stocks, 18".

Centre Rest will take work to 3⅛" in diameter.

Weight, about 350 lbs.

Table, index plates and tables explaining the use of same, wrenches and everything shown in cut are sent with each pair of centres.

8 in. and 6 7-8 in. SINGLE DIAL
INDEX CENTRES.

These Index Centres are intended for use on milling or other machines where a limited amount of indexing is to be done, as in cutting teeth in sprocket wheels or mills, or in milling nuts, etc.

Two sizes are made, one taking in work to 8″ in diameter and the other to 6⅞″ in diameter.

The Spindles are threaded on the ends. The 8″ spindle has a No. 10 taper hole and the 6⅞″ spindle a No. 11 taper hole. The 6⅞″ centres are especially designed for use in milling sprocket wheels in gangs and provision is made for a stiff arbor.

The Index Plate is a dial provided with hardened steel bushings, and is covered, thus protecting the holes from dirt. A hardened steel taper pin is forced into the bushing by a spring, and can be released by a lever, when the work can be rotated by a hand wheel, thus making the indexing very rapid. While the plates can be used, usually, for other than the number of teeth for which they are made, it is desirable to have them contain holes for the number of teeth to be cut, as mistakes can thus be avoided.

The Dial furnished with the centres has 16 holes. Special dials, with any number of holes to 25, made to order.

The Tongues and Bolts furnished, fit a T slot ⅝″ wide.

Combined Length of head and foot-stocks, 17½″.

Wrench, bolts and clamps are furnished with these centres.

10 INCH INDEX CENTRES.

These centres enable one to do all of the dividing and indexing work that can be done on a Universal Milling Machine, excepting the cutting of spirals, or of work that has to be held at an angle to the centre line of the centres.

The Centres swing $10\frac{1}{4}''$ in diameter.

The Spindle is threaded on front and has a No. 10 taper hole. The straight hole at end of taper $1\frac{1}{16}''$ in diameter.

The Worm Wheel is $6\frac{1}{2}''$ in diameter, and one revolution is made by 40 revolutions of index crank. It has 24 holes on rim, and when the worm is disengaged direct indexing can be done. The wheel is held by means of an index pin.

The Index Plates are the same that are used on the No. 1 Universal Milling Machine.

The Head-Stock can be clamped at any angle on table.

The Tongues and Bolts furnished, fit a T slot $\frac{5}{8}''$ wide. The tongues are inserted.

Combined Length of head and foot-stocks, $13\frac{3}{4}''$.

---

12 INCH INDEX CENTRES.

These Centres are of the same general design as the 10″ Index Centres described above.

The Centres swing $12\frac{1}{4}''$ in diameter.

The Worm Wheel is $7\frac{3}{4}''$ in diameter.

The Tongues and Bolts furnished, fit a T slot $\frac{3}{4}''$ wide.

Combined Length of head and foot-stocks, $16\frac{3}{4}''$.

Index Plates and tables explaining the use of same and wrenches are sent with each pair of Centres.

TOOL-MAKERS' UNIVERSAL VISE.

This vise is of an entirely new design, for use on Milling Machines or Planers, and is so constructed that it can be set at any angle to the surface of the table or to the spindle of the machine, and rigidly clamped in position.

The base is double, the upper portion is graduated, and can be set at any angle in a horizontal plane. On top of the swivel base is a hinged knee, which can be set at any angle, to 90°, in a vertical plane. The top of the knee is graduated. The knee is clamped rigidly in position by means of the nut on end of bolt forming hinge, and the locking levers shown at the left of cut, these levers are clamped in position by the bolt shown in centre and the bolts at the ends of the levers.

The vise proper is fastened to the hinged knee in such a manner that it can be set at any angle on a horizontal plane, and can be clamped in position by the bolt which holds the upper locking lever.

The vise base is fastened to the table by means of two bolts fitting into the table T slot. The base is provided with two sets of holes to allow for moving the vise, when set in a vertical plane, in order to clear the Milling Machine spindle.

The jaws are made of tool steel and hardened, are 5⅛" wide, 1¼" deep and will open 2¾".

MILLING CUTTER.

MATERIAL, WORKMANSHIP, DESIGN, ETC.

Our machines are made with the intention that they shall be the best of their respective classes. Materials are used which experience has shown are the most suitable for the various parts, and careful attention is constantly given to insure good workmanship.

Our buildings are modern and specially arranged for the business. The machinery and appliances are the best attainable. Our machines are manufactured in large quantities with expensive special tools, and much greater accuracy has been obtained than can be reached by the usual methods of manufacturing. All plane bearings are scraped to surface plates; all cylindrical bearings are ground and fitted to standards. The alignments are correct. The machines are subjected to thorough inspection and when deemed necessary, to actual operation before being packed.

The general design of our Universal Milling Machines has, for thirty years, been appreciated by mechanical experts, and faithfully imitated by almost every one who has desired to enter upon the manufacture of Milling Machines. The design of the Plain Machines, in the main features, is similar to that of the Universal Machines. The distinguishing features of the Nos. 12 and 13 Plain Milling Machines are the rapidity with which they can be operated; and their convenience for setting and removing the work. They are also unusually stiff for their size and weight.

We are constantly endeavoring to improve the machines, and from time to time modify the details, and for this reason we suggest that our catalogue be frequently consulted, as all changes are there most speedily brought to the attention of the public.

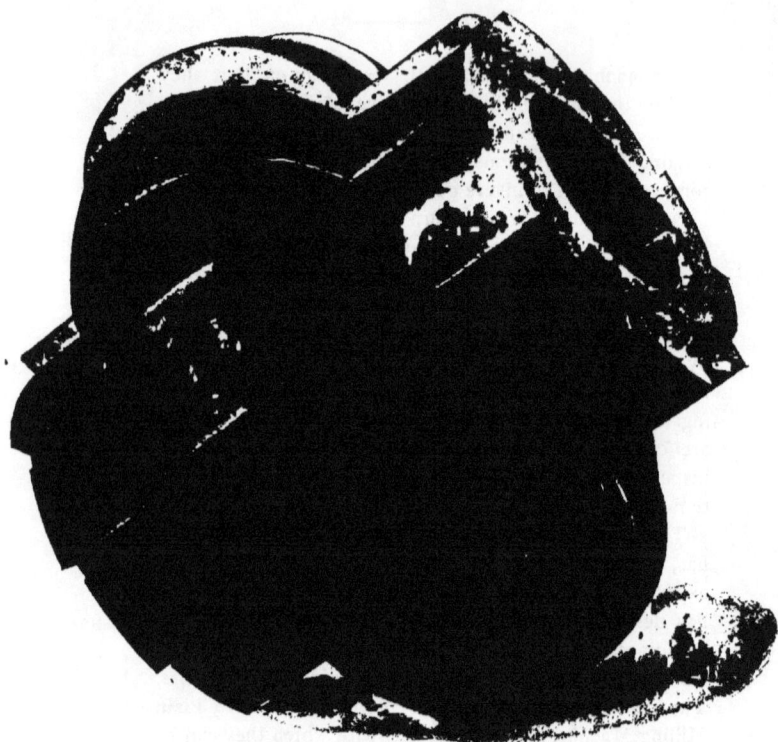

FORMED CUTTER.

CARE AND USE OF MILLING MACHINES.

The Machines should be placed upon a level, and if possible, a solid floor or foundation. The Universal and several of the Plain Machines should not be set where the screw which raises and lowers the knee will come directly over a beam : for there must be room for this screw below the base of the machine in order that the knee may reach its lowest position. *Placing Machines.*

The counter-shafts are generally placed over the machines, but if necessary the position of the cone and other pulleys may be changed. *Placing Counter-shafts.*

The shippers are most convenient when on the left side of the machines.

To find the diameter of pulley required to run the counter-shaft at a given speed : multiply the speed of counter-shaft by the diameter of pulley on the same, and divide the product by the speed of main, or driving shaft. *Diameter Pulley on Main Line*

It is sometimes best to vary the speed of counter-shafts from those given in our catalogue. In our own works some counter-shafts belonging to No. 1 Universal Milling Machines are run at only 90 turns per minute, while others are run at 150 turns per minute. *Speed of Counter-shafts.*

As the life and efficiency of machines depend largely upon the amount of care bestowed upon them, it is important that they should be kept clean and well oiled, and that all repairs should be promptly made.

A coal or mineral oil called No. 2 cosmoline, we have found to be excellent for lubricating. In oiling the Universal Milling Machines the screw in the table should be replaced to keep the dirt out of the clutch gear bearings and the feed screw. These are oiled through the table oil hole where the lines on the table and bed match. Several oil holes are marked and have threaded stops. *Oiling.*

FORMED CUTTER.

The back gear of the No. 4, No. 3 prior to 1893, Universal Milling Machine is oiled through the screw hole W, Fig. 17, and enough oil is held in the quill to last for several months. There are two other oil holes in the cone of this machine.

The greatest care should be taken that chips do not get into the holes in the spindles or between the arbor collars. Chips should also be kept from between the knee and frame.

In placing the tools belonging to the Universal Milling *Placing Tools in the Closets of Machines.* Machines in the closets, the following method is convenient: Put the wrenches, centre keys, collet and spanner upon the top shelf; the chuck, centre rest and raising block upon the second shelf from the top; the change gears and index plates in the wooden compartment, and the vise and its clamps upon the bottom shelf. When the hand wheel is not in use it should hang on the stud at the side of the machine, and when the machine is not in use it is well to have the spiral head and footstock on the table.

For catching oil from the work it is convenient to have *Oil Pans.* two pans, one about 5¼ inches square, and the other 5¼ inches wide and 8 inches long. The pans can be about 1¼ inches deep and should have a strainer near the top. Cleats on the bottom of the pans will keep them from slipping off the table. These pans are not sent with the machines, but may be ordered from us if desired.

The construction of the machine should be examined *Adjusting Spindle.* carefully before any part is removed or adjusted. The proper adjustment will allow the spindle to be easily revolved by hand. The adjustment of the spindles will not need to be changed for a long time after the machines have left our works.

The question is often asked, How much skill is required to properly use or operate milling machines? A conservative answer is given in the following editorial from the *American Machinist:*

METAL SLITTING SAWS AND SLOTTING CUTTERS.

MILLING MACHINES AND SKILL.

"No one who has had sufficient experience with the milling machine in its various forms to acquire a reasonably clear idea of its capabilities, and who has an opportunity to see the machine in use in the various shops, can fail to see that in many of them it is very imperfectly understood, and that, as a consequence, comparatively poor results are obtained from its use—results, we mean, which are very poor compared with those which should be obtained, and are obtained in every case where the legitimate functions of the machine are clearly recognized, and the conditions necessary to its successful operation secured.

The milling machine intelligently selected or constructed, with reference to the work it is expected to do, provided with well-designed and well-made special fixtures, where the nature of the work calls for them, and then skillfully handled, is a surprisingly efficient tool, but used as it is being used in many shops to-day, it is a delusion, a failure, and an injury alike to the users, to the builders, and to the good name of milling machines generally.

While it is true that there is scarcely a machine tool in use which will yield more satisfactory returns for a given outlay, when pains are taken to use it in the best possible manner, it is also true, we think, that there is no tool in common use, the efficiency of which is so much reduced by careless or ignorant handling and abuse. Considerable intelligence and skill, as well as constant attention, must be bestowed upon the milling machines in order to secure anything like a satisfactory performance from them, either in the quality of the work done or in its quantity.

In some cases this skill and intelligence must be possessed and exercised by the man who actually handles the machine, in other cases by some one who, though he

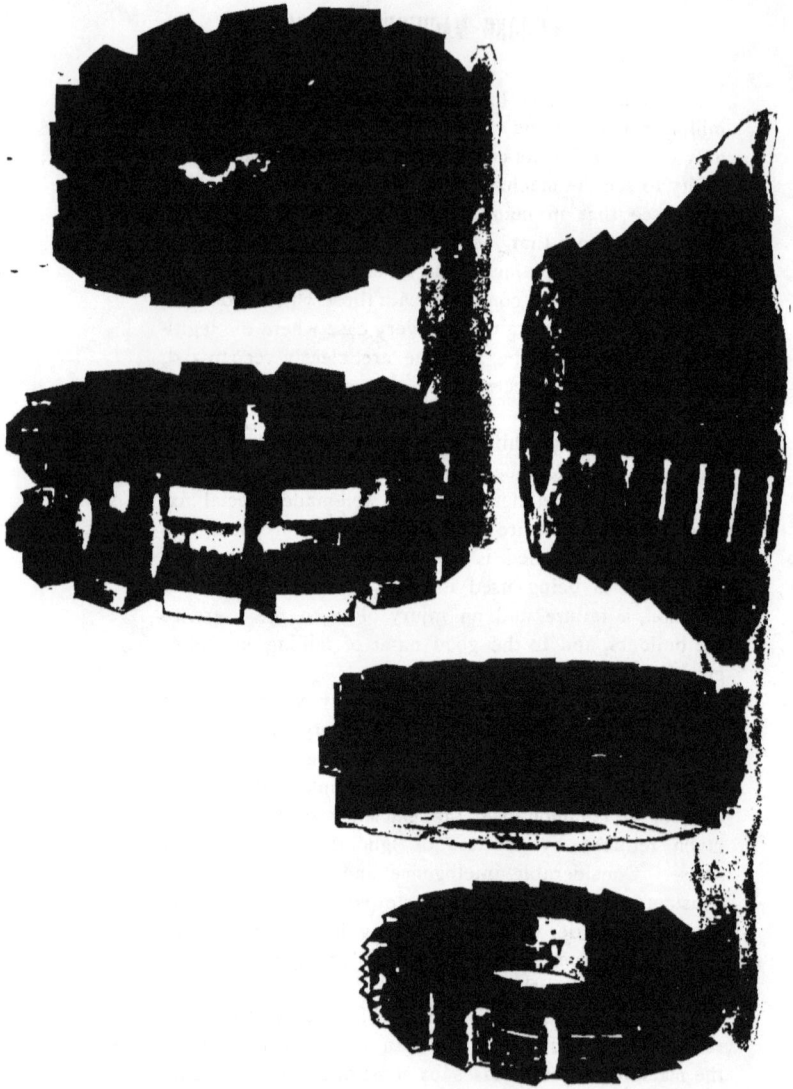

FORMED CUTTERS.

does not actually operate it, supervises its operation, and is responsible for the work done by it. But in any case, the skill, the intelligence and the careful attention must be exercised, or the results will be anything but satisfactory.

We hear a great deal about the comparatively cheap labor required to do milling machine work, and it is evident that too many shop proprietors have concluded from this that about all that is necessary to do such work is to buy the machine, hire a boy to run it, have him " shown how" for an hour or so by one of the lathe hands, and then let the boy and the machine work out their own salvation.

No greater mistake could possibly be made, and it is in such a shop that a milling machine man finds the machine working often at less than half its capacity, with an apology for a cutter, ground by hand in every shape but the right one, two or three only of its superabundant teeth touching the work, and they, with a distinct thump and knock, indicating anything but a real cutting action, while the boy stands by and occasionally— when it occurs to him to do so — squirting a few drops of black lubricating oil onto the chips with which the spaces between the teeth are tightly jammed. The proprietors of such shops are not usually very enthusiastic regarding the use of milling machines, and it would be a wonder if they were.

Where a universal milling machine is used upon tool work, or for other purposes requiring a constant change from one job to another, it is a mistake to suppose that there is economy in the employment of a boy or cheap man to operate it. And many of those who think they are saving money in that way would be greatly surprised to see the work turned out from such machines by good mechanics who thoroughly understand them, and are capable of earning good wages upon them.

Experience has proved that it pays as well to put first-class mechanics upon such machines as upon any other machine tools.

Where milling machines are used for regular manufacturing operations, and the same cycle of movements is to
be repeated for a large number of pieces, boys, or men
who are not skilled mechanics, answer every purpose;
but the skilled supervision must be there, and it must be
seen to that the machine is as well taken care of, the
cutters as well made and ground, and in fact, everything
as well done as though a good mechanic actually operated
the machines.   In fact, in the shops in which the best
results are obtained from the use of milling machines in
regular manufacturing operations all changes of the
machines from one job to another, all adjustments, and
the grinding and replacing of cutters are done by, or, at
least, under the direct supervision of a skilled mechanic,
responsible for the work of the machines, and who thoroughly understands and appreciates them.   In this way
only can the full benefits of the machine be realized.

It is far too common to go into the tool-room and find
a splendid universal milling machine standing idle, while
perhaps two or three men are doing at the shaper, planer
or vise, jobs which could be done by an expert milling
machine hand, in one-fourth down to one-tenth of the
time, and a great deal better.   One fault, which is far
more common than would readily be believed in some
quarters, is a failure to recognize the fact that a milling
machine necessarily calls for some sort of machine for
grinding cutters, and that a machine upon which cutters
are used, that are ground by holding the edges one after
the other against an emery wheel by hand, is at a
decided disadvantage, and will do no work which either
in quality or cost will make a favorable showing when
compared with that which is done by properly ground
cutters.

It should be much more generally recognized that in
milling machine practice, as in other things, there is a
right way and a wrong way, and that skilled, intelligent
labor pays best.   When these facts are more generally
recognized, it will be better for both the builders and for
the users of the machines."

FORMED CUTTERS.

CUTTERS USED ON MILLING MACHINES.

The most simple of the form cutters is the fly cutter shown with its holder in Fig. 20, the cutting face as shown by the end view being held about in line with the centre of the holder. As these cutters have but one cutting edge they mill accurately to their own shape, but of course do not mill so fast, or wear as long as cutters with a number of teeth. They can be formed very exactly to any desired shape at comparatively small expense, and thus may be used for many operations that otherwise will not bear the cost of special cutters;— for example: when one or two small gears are wanted in experimental work. Fly cutters are also of special advantage in making and duplicating screw machine and other tools of irregular cutting contour. The clearance in tools thus made may be obtained by holding the tool blank in the vise so that the front end will be elevated several degrees.

Formed Mills or Formed Cutters

As used by us the term "Formed Cutters" applies to the cutters with teeth so relieved that they can be sharpened by grinding without changing their form, while "Form Cutter" can be applied to any cutter cutting a form, regardless of the manner in which the teeth may be relieved. Fig. 25 represents a formed cutter, Fig. 26 a form cutter.

Single formed mills, as shown in Fig. 21, are not uncommonly made in one piece 7 inches diameter or 6 inches long. When the width of the cut is greater than can be easily made by one cutter several cutters are combined in a gang as in Fig. 25. A gang is limited in length only by the capacity and power of the milling machine.

Many users of Milling Machines are not fully aware of the variety of Cutters that are made and carried in stock.

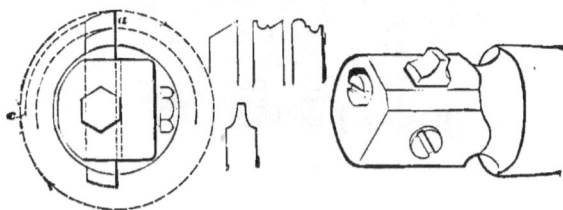

FIG. 20.

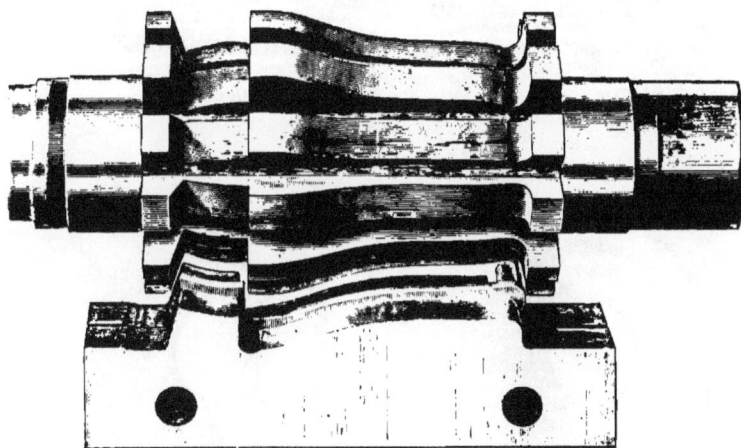

FIG. 21.

Milling Cutter.

Left Hand End Mill.

End Mill with Centre Cut.

Side Milling Cutter.

Screw Slotting Cutter.

Metal Slitting Saw.

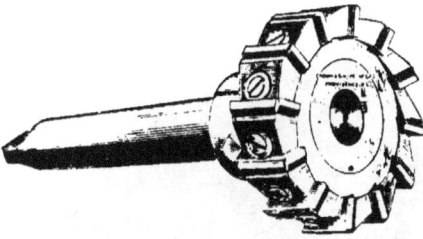

End Mill with Inserted Teeth.

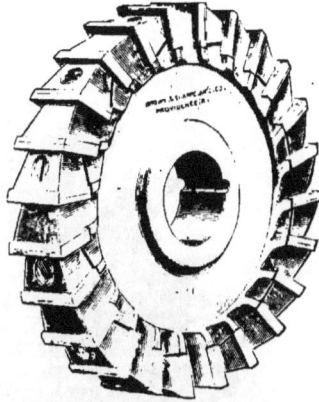

Side Milling Cutter with Inserted Teeth.

Involute Gear Cutter.

Angular Cutter.

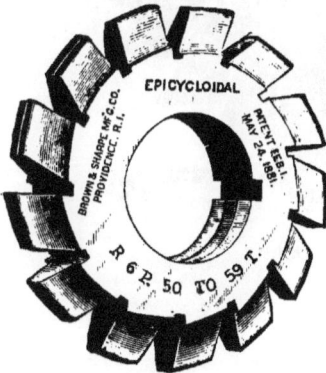

Epicycloidal Gear Cutter.

Stocking Cutter.

SIDE MILLING CUTTER.

We now manufacture twenty-six varieties and ten hundred and thirty-eight sizes of stock cutters, and we can make any size or shape, or arrange for any combination of cutters that may be desired. The formed cutters can be sharpened by grinding without changing their outline.

On this page we give outline cuts showing the forms cut by the respective cutters, and on other pages we give full page cuts of stock and special cutters.

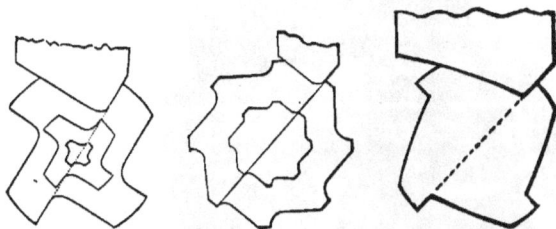

Tap and Reamer Cutters.

Four Lipped Twist
Drill Cutter.

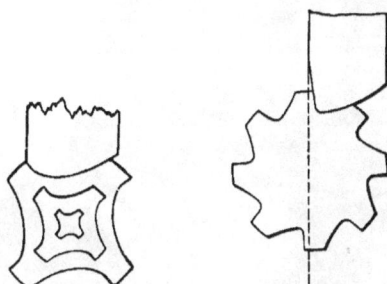

Tap Cutter.

Reamer Cutter.

It is well to have mills or cutters as small in diameter *Diameter of Mills.* as the work or their strength will admit. The reason is shown by Fig. 22. Suppose the piece I D C J E is to be cut from I J to D E. If the large mill A is used, it will strike the piece first at I when its centre is at K, and will finish its cut when the centre is at M. The line G shows how far the mill must travel to cut off the stock

GEAR CUTTERS.

I J D E.   If the small mill B is used, however, it travels only the length of the line H.   It can also be seen that a tooth of B travels through a shorter distance between the lines D E and I J than a tooth of A.   This is true of all ordinary work, or where the depth of cut I D is not more than half the diameter of the small mill.

The advantage of small mills has been illustrated in our own works, where a difference of ½ an inch in the mills has made a difference of 10% in the cost of the work.

In short, small mills do more and better work, cut more easily, keep sharp longer and cost less than large mills.

When it is possible the mill should be wider than the work, and the hole in a mill should be as small as the strength of the arbor will admit.   The stock around the hole, however, should not be less than ⅜ of an inch thick. *Length of Mills. Diameter of Hole in Mills.*

A mill is not necessarily too soft because it can be scratched with a file, for sometimes when cutters are too hard or brittle and trouble is caused by pieces breaking out of the teeth they can be made to stand well and do good work by starting the temper. *Temper of Mills.*

Of late years mills have been made with coarser teeth than formerly, the advantages being more room for the chips and less friction between the teeth and the work. When the teeth are so fine that the mill drags, or the stock is powdered the mill heats quickly and does not cut freely. *Number of Teeth of Mills.*

The friction may also be reduced, especially in large mills taking heavy cuts, by nicking or cutting away parts of the teeth, which breaks the chips and allows heavier cuts and feeds to be taken.

Knowing the conditions under which a mill is to be used in our own practice, we modify the number of teeth as seems expedient, usually making the special mills coarser in pitch than the stock mills, for our observation indicates there are more mills with too many teeth than with too few.   But sometimes we relatively increase the

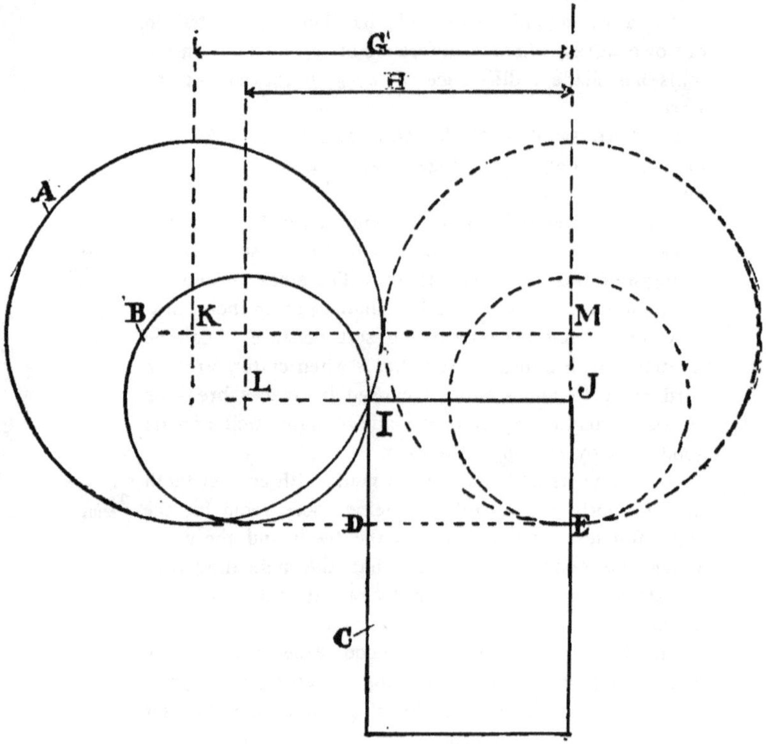

FIG. 22.

number of teeth, as for instance, large mills, in some cases, ·can advantageously be designed to have more than one tooth cutting all the time on broad surfaces and in deep cuts.

In England cutters generally have finer teeth than in America, the pitch being about two-thirds as much as ours, as shown by the rules given at a meeting of the Institution of Mechanical Engineers in London, October 30, 1890. These rules we are permitted to print by the courtesy of the gentlemen who presented them at the meeting.

Mr. Geo. Addy, of Sheffield, estimates the pitch of teeth of cutters from 4 inches to 15 inches diameter by the following rule :

Pitch in inches $= \sqrt{\text{(diameter in inches} \times 8)} \times 0.0625$. Mr. J. Macfarlane Gray pointed out that this rule might be put into a somewhat simpler form for more convenient use as a plain workshop rule, by saying that the product of the pitch in inches multiplied by the pitch in thirty-seconds of an inch was 'equal to the diameter in inches, and that a still more simple way of stating what was substantially the same rule was to say that the number of teeth in a milling cutter ought to be one hundred times the pitch in inches ; that is, if there were 27 teeth, the pitch ought to be 0.27 inch.

In regard to the cutting angle of the teeth we in theory agree with what Mr. Addy stated in the paper before referred to, viz. : " The adoption of the most suitable cutting angle should receive the same close attention that is now universally bestowed upon the ordinary tools for turning and planing." But in practice while in many instances adopting the angle according to the material to be used, yet taking into consideration all the conditions of using and caring for the cutters, we have generally found it satisfactory to have the cutting edges of the teeth radial. As a result of considerable research and experience Mr. Addy gives as his opinion that the front of the teeth instead of being truly radial should have a backward

Cutting Angle of Teeth of Mills.

inclination of 10 degrees from the radius, the cutting angle in this way being 70 degrees and the clearance angle 10 degrees.

Clearance of Mills. The relief or clearance of mills we think should usually be about three degrees, and the land at the top of the teeth from .02 inches to .04 inches wide before the clearance is cut or ground.

On cotter mills, page 135, the clearance should be at the outside of the two teeth, as the cutting is not done directly upon the end; but this clearance should usually be only about one-quarter of that on other mills.

Mills to cut grooves should be hollowing about five one-hundreths in one inch for clearance, that is, a grooving mill should be about one one-hundreth of an inch thinner at one inch from its edge or circumference than it is at the edge. Our grooving mills are given a limit of two one-thousandths in thickness. Mills made to exact thickness are very expensive. In cutting grooves that are to have some parts of their sides nearly or quite parallel, it is well to leave considerable stock, for the finishing cut, as mills like taps do better work when they can get well into the stock.

For the same reason if the sides have to be left slanting before the finishing cut is taken, no part of either side should slant less than three degrees, or about one $\frac{1}{20}$ of an inch in each inch.

Clearance in a fly cutter is obtained by moving the cutter towards a, Fig. 20.

Sketches of Mills. A sketch of a mill with a solid shank, or of a formed cutter that has half of its outline unlike the other half, should clearly show which way the mill is to turn. This can be done by an arrow, as shown on page 135, or by writing the word "coming" either at the top or bottom of the sketch, as the case may be. This cut also shows how the terms right and left are applied to angular cutters and mills. It is well to show the work in red lines in the position that it will occupy when cut by the mill.

COTTER MILL.

L. H. THREAD.   R. H. THREAD.

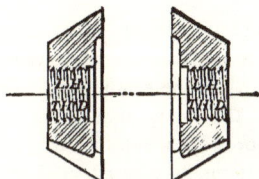   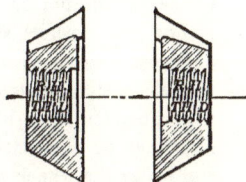

COMING   *GOING*
LEFT HAND   RIGHT HAND   *RIGHT HAND   LEFT HAND*

ANGULAR CUTTERS.

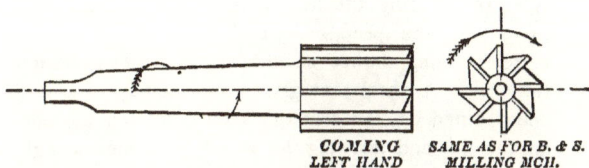

COMING   *SAME AS FOR B. & S.*
LEFT HAND   *MILLING MCH.*

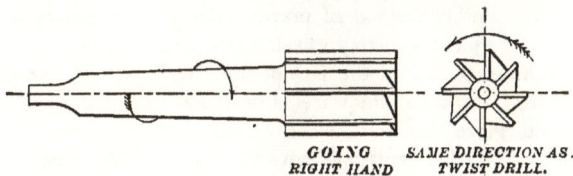

GOING   *SAME DIRECTION AS A*
RIGHT HAND   *TWIST DRILL.*

END MILLS.

Templets. In ordering formed cutters it is well to send templets of the desired shapes, and to state how nearly exact the cutters must be. If any part of the outline is to be a straight line the sketch should clearly show this, and if any part is to be circular the radius should be given. Unless instructions have been given to the contrary, mills are generally made and hardened to cut steel or iron.

Sharpening Mills. A dull mill wears away rapidly and does poor work. Accordingly care must be taken to keep mills sharp. In sharpening them it is necessary to be very careful that the temper should not be drawn.

Grade of Emery Wheel required for the work. The emery wheel should be of the proper grade as to hardness and as to the size of the emery. The wheel should be soft enough so that it can be easily scratched with a pocket knife blade, and the emery should not be finer than 90 nor coarser than 60. As a rule, the coarser and softer the wheel, the faster it should run, although the periphery speed should not exceed 5000 feet per minute.

Width of Face of Wheel. A wheel of the proper grade should be used with the face not to exceed ¼ " wide. If the wheel glazes, the temper of the cutter will be drawn. In such a case, if the wheel is not altogether too hard, it can sometimes be remedied by reducing the face of the wheel to about ⅛" or by reducing the speed, or by both.

Before using, a wheel should be turned off so that it will run true. A wheel that glazes immediately after it has been turned off can sometimes be corrected by loosening the nut and allowing the wheel to assume a slightly different position when it is again tightened.

Another method of preventing a wheel's glazing is to use a piece of emery wheel, a few grades harder than the wheel in use, on the face of the wheel, whereby the cutting surface of the wheel is made more open and less apt to glaze.

Take light cuts and move the cutter rapidly across the face of the wheel. In one of our circulars is published a list of wheels and speeds suitable for tool grinding machines.

Mills that have their teeth ground for clearance are particularly apt to have their temper drawn in sharpening, especially at the edge of the teeth, and often when the temper has been drawn and the teeth are polished, they will look as usual after being ground.

In sharpening angular cutters on the face, it is best to leave the side of the teeth crowning or a little higher toward the centre or hole of the cutter than towards the point of the teeth. Sharpening Angular Cutters.

Formed cutters are sharpened by central or radial grinding upon the front of the teeth, square across the cutter or in line with the axis, for if the teeth are not ground radially, the work done by them will not be of the correct shape. The tendency, however, in grinding these mills is to take away too much from the outer part of the front of the teeth. An attachment for grinding formed cutters is made for our No. 3 Universal Cutter and Reamer Grinder. Sharpening Formed Cutters.

The best plan is to have all mills sharpened immediately after they have been used, before they are put away.

The advantage of properly sharpening cutters is indicated by the amount of work done by the gear cutter shown on page 138. This cutter was, when new, the same in appearance as the gear cutters shown on page 127, and it has cut 467, 4 pitch, 64 teeth, 3-inch face cast iron gears,—making a total length of cut of 7472 feet. The teeth of the gears were cut from solid blanks and finished in one cut. This record while good is not exceptional.

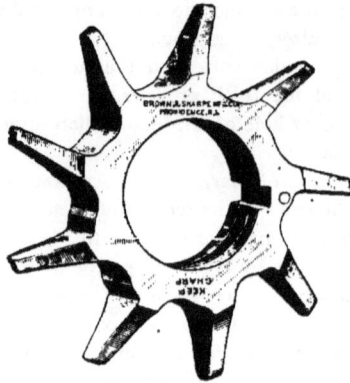

GEAR CUTTER WORN.

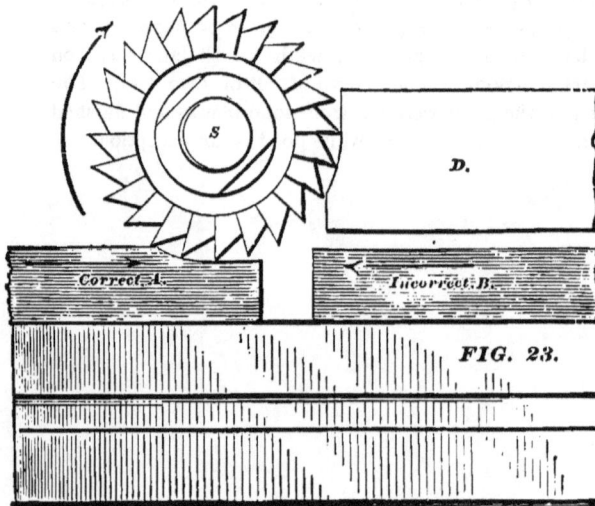

FIG. 23.

GANG OF CUTTERS.

USE OF MILLING MACHINES.

EXAMPLES OF OPERATIONS.

Use of Oil on the Cutters or Work. Oil is used in milling to obtain smoother work, to make the mills last longer, and, where the nature of the work requires, to wash the chips from the work or from the teeth of the cutters. It is generally used in milling a large number of pieces of steel, wrought iron, malleable iron or tough bronze. When only a few pieces are to be milled it frequently is not used, and some steel castings are milled without oil; also in cutting cast iron it is not used. For light, flat cuts it is put on the cutter with a brush, giving the work a thin covering like a varnish; for heavy cuts it should be led to the mill from the drip can, sent with each machine, or it should be pumped upon or across the mill in cutting deep grooves, in milling several grooves at one time, or indeed, in milling any work where, if the chips should stick, they might catch between the teeth and sides of the groove and scratch or bend the work.

Generally we use lard oil in milling, but any animal or fish oils may be used. The oil may be separated from the chips by a centrifugal separator, or by the wet process, so that a large amount may be used with but little waste.

Some manufacturers prefer to mix mineral oil with lard or fish oil, and state the mixture is less expensive and works well. Prof. J. E. Denton has made experiments with mixtures and thinks that mineral or coal oil can be advantageously used.

An excellent lubricant to use with a pump is by mixing together and boiling for one-half hour, ¼ pound Sal Soda, ½ pint Lard Oil, ½ pint Soft Soap and water enough to make 10 quarts.

There is a difference of opinion as to whether the work should be moved against the cutter as at A, Fig. 23, or with it as at B. But in most cases our experience and experiments show it is best for the work to move against the mill as shown at A, Fig. 23.

When it moves in this way the teeth of the cutter, in commencing their work, as soon as the hard surface or scale is once broken, are immediately brought in contact with the softer metal, and when the scale is reached it is pried or broken off. Also when a piece moves in this way, the cutter cannot dig into the work as it is liable to do when the bed is moved in the direction indicated at B. When a piece is on the side of a cutter that is moving downwards, the piece should, as a rule, have a rigid support and be fed by raising the knee of the machine.

Direction in which Work is Moved under a Mill.

Some work, however, is better milled by moving with the cutter. For example : To dress both sides of a thick piece D with a pair of large straddle mills, it might be well to move the piece towards the left, as the mills then tend to keep it down in place instead of lifting it.

Again in milling deep slots, or in cutting off stock, with a thin cutter or saw, it may be better to move the work with the cutter, as the cutter is then less likely to crowd side-wise and make a crooked slot.

When the work is moving with the cutter, the table gib screws must be set up rather hard, for if the work moves too easily the cutter may catch and the cutter or work be injured. A counter weight to hold back the table is excellent in such milling.

For the purpose of making a comparative test of the two methods, we made the following four experiments on our No. 5 Plain Milling Machine :

Experiments to test method of running Cutter.

This machine had been provided with a take-up attachment for back-lash of table, two cutters of same diameter and width were used, and suitable castings were provided, same being 3″ square and 3 ft. long (pickled).

First experiment with No. 1 Cutter was with the cut, cutting down on scale and feeding 6″ per minute. After cutting one surface 3 feet long, the cutter was found to be dull.

Second experiment with No. 2 Cutter was against the cut, cutting under the scale and feeding 6″ per minute. Eight castings 3 feet long were milled before the cutter had shown the wear of No. 1.

In order to prove the tempering of the Cutters both were re-ground.

Third experiment was with No. 1 Cutter working in same manner that No. 2 Cutter had been, viz. : against the cut, under the scale and feeding 6″ per minute. Eleven castings 3 feet long were milled before the Cutter had shown the wear of No. 2.

Fourth experiment was with No. 2 Cutter working in the same manner as No. 1, as described in first experiment. This Cutter failed on the first cut.

Speed of Mills.

It is impossible to give definite rules for the speed and feed of Cutters, and what is here said is only in the way of suggestions. Sometimes the speed must be reduced, and yet the feed need not be changed. The judgment of the foreman or man in charge of the machine should determine what is best in each instance.

Average Speed.

The average speed on wrought iron and annealed steel is perhaps forty feet a minute, which gives about sixty turns a minute for Cutters 2½ inches diameter. The feed of the work for this surface speed of the Cutter can be about 1½ inches a minute, and the depth of cut say $\frac{1}{16}$ of an inch. In cast iron a Cutter can have a surface speed of about fifty feet a minute while the feed is 1½ inches a minute and the cut $\frac{3}{16}$ of an inch deep, and in tough brass the speed may be eighty feet, the feed as before and the chip $\frac{3}{32}$ of an inch.

As a small Cutter cuts faster than a large one, an end mill for example, ½ inch diameter can be run about 400 revolutions with a feed of 4 inches a minute.

For examples of what may regularly be done under suitable conditions, we may mention that Cutters 2½ inches in diameter used in cutting annealed cast iron in our works are run at more than 200 turns, or at a surface speed of more than 125 feet, while the work is fed more than eight inches a minute. The cuts are light, not more than $\frac{1}{32}$ of an inch deep, and the work is short, from $\frac{1}{2}$ inch to 1 inch long. Two side mills 5 inches in diameter running 50 turns a minute, dress both edges of cast iron bars $\frac{3}{4}$ of an inch thick, with a feed of more than 4 inches a minute.

An English authority, Mr. Geo. Addy, gives as safe speeds for cutters of 6 inches diameter and upwards :

Steel, · 36 ft. per minute with a feed of $\frac{1}{4}''$ per minute.
Wrought Iron, 48 " " " " $1''$ "
Cast Iron, 60 " " " " $1\frac{1}{2}$ "
Brass, 120 " " " " $2\frac{1}{4}$ "

And he gives as a simple rule for obtaining the speed :— Number of revolutions which the cutter spindle should make when working on cast iron $= 240$ divided by the diameter of the cutter in inches.

Mr. John H. Briggs, another English authority, states, "for cutting wrought iron with a milling cutter taking a cut of one inch depth — which was a different thing from mere surface cutting — a circumferential speed of from 36 to 40 feet per minute was the highest that could be attained with due consideration to economy, and to the time occupied in grinding and changing cutters ; the feed would be at the rate of $\frac{5}{8}$ inch per minute. Upon soft mild steel, about 30 feet per minute was the highest speed, with $\frac{1}{4}$ inch depth of cut and $\frac{3}{4}$ inch feed per minute. Upon tough gun-metal, 80 feet per minute, with $\frac{1}{2}$ inch depth of cut and $\frac{3}{4}$ inch feed. For cutting cast iron geared wheels from blanks previously turned, and using in this case comparatively small milling cutters of only $3\frac{1}{2}$ inches diameter, the speed was $26\frac{1}{2}$ feet per minute, with $\frac{1}{2}$ inch depth of cut and $\frac{3}{4}$ inch feed per minute."

Slotting cutters may often be run at a higher speed than other cutters of the same diameter, but with a wider face. Angular cutters must in some instances be used with a fine feed to prevent breaking the points of the teeth.

The following table may be of service as suggesting speeds that may be tried with our machines on ordinary work, but it is not published as an absolute guide, and as before stated, the judgment of the foreman must determine what is best in each instance. In considering the table it must be borne in mind that rapid progress is being made in milling, and that all figures are submitted with the certainty that improvements in machines, cutters and fixtures will soon render them obsolete.

Limits in Milling. An ordinary limit is four one-thousandths of an inch. This is allowable for bolt heads, nuts, and the squares at the ends of shafts where cranks or hand wheels are used, also for some kinds of gibs and many parts that are milled for a finish.

In most sewing machine pieces, electrical and scientific instruments, type writers and fine machinery, the limit is two one thousandths. Thus a slot that is called half an inch wide may be any size between half an inch and five hundred and two one thousandths of an inch (.500" to .502") while the tongue or piece that goes into the half inch slot may be of any size between two one-thousandths less than one half inch and one-half inch (.498" to .500"). On many pieces, for instance usually on those milled for a finish, the limit may of course be either above or below the standard size.

Some work should be milled as close as possible to exact size; and when close fits are required it is often cheaper and better to do the fitting by the milling machine than by filing or other hand work.

The most accurate results in milling to a given thickness or size are ordinarily obtained by straddle mills or side milling cutters; for when only one side is milled at a time and the piece has to be changed from side to side, it is hardly practicable to work to a smaller limit than two one-thousandths of an inch. Side milling frequently requires more attention to keep the work smooth than ordinary sur-

SOFT MACHINERY STEEL.

Diam. of Mill.	Rev. per Minute.	Speed of Cutter pr. Minute.	Depth of Cut.	Width of Cut.	Feed per Minute	
					In Scale of Steel.	Under Scale of Steel.
$1\frac{1}{2}''$	240	32'	1-16''	1-2''	3 3-4''	4''
	240	32	9-32	1-2	3-4	1
1	130	34	3-32	1	2 1-4	3
	130	34	3-8	1	1-2	3-4
$1\frac{3}{4}$	54	25	1-16	1 3-4	2 3-4	3
	54	25	5-8	1 3-4	3-8	1-2

SURFACE MILLING SOFT MACHINERY STEEL.

Diam. of Mill.	Rev. per Minute.	Speed of Cutter pr. Minute.	Depth of Cut.	Width of Cut.	Feed per Minute	
					In Scale of Steel.	Under Scale of Steel.
3	46	37	1-16	1	4	6
	46	37	9-16	1	1-4	3-8
	46	37	1-16	2	1 1-2	2
	46	37	9-16	2	3-16	1-4
	46	37	1-16	3	1	1 1-2
	46	37	7-16	3	1-4	5-16
$4\frac{1}{2}$	36	42	1-16	3	1 1-4	1 1-2
	36	42	7-16	3	1-4	5-16
	36	42	1-16	6	8-4	1
	36	42	7-16	6	1-8	3-16

CAST IRON.

Diam. of Mill.	Rev. per Minute.	Speed of Cutter pr. Minute.	Depth of Cut.	Width of Cut.	Feed per Minute	
					In Scale of Cast Iron.	Under Scale of Cast Iron.
$1\frac{1}{2}''$	305	40'	1-32''	1-2''	21''	88''
	305	40	3-32	1-2	3	8
$\frac{7}{8}$	174	40	3-32	7-8	7	10 1-4
	174	40	9-16	7-8	7-8	1 1-8
$1\frac{3}{4}$	130	60	1-8	1 3-4	3	4 1-2
	65	30	13-16	1 3-4	7-8	1 1-2

SURFACE MILLING CAST IRON.

Diam. of Mill.	Rev. per Minute.	Speed of Cutter pr. Minute.	Depth of Cut.	Width of Cut.	Feed per Minute	
					In Scale of Cast Iron.	Under Scale of Cast Iron.
3	46	37	1-8	1	5 1-2	7
	46	37	1-8	2	3 1-8	5
	46	37	1-8	3	2	3 1-2
	46	37	1-2	1	1 3-4	2
	46	37	1-2	2	1-4	1 1-2
	46	37	1-2	3	3-4	1
$4\frac{1}{2}$	36	42	1-16	3	6 1-2	11
	36	42	7-16	3	1-2	5-8
	36	42	1-16	4	3 1-2	5
	36	42	7-16	4	2	3
	36	42	1-16	6	1-8	4
5	30	35	13-16	6	10	3-16
	20	35	1-16	5	2	12
	20	35	1-4	5	2	3

SPEEDS AND FEEDS OF No. 4 PLAIN MILLING MACHINE.

END OR FACE MILLING CAST IRON.

Diam. of Mill.	Revs. per Minute.	Speed of Cutter pr. Minute.	Depth of Cut.	Width of Cut.	In Scale of Cast Iron.	Under Scale of Cast Iron.
1/2"	310	40'	1.64"	1-2"	10 3-8"	72"
	310	40	3-16	1-2	3 3-8	8 1-4
7/8	200	60	1-16	7-8	25 1-2	63
	200	60	5-8	7-8	7	10 3-4
1 1/4	170	80	1-16	1 3-4	25	40
	170	80	3-4	1 3-4	1 1-8	1 7-8
5	24	30	13-32	5	2 3-8	3 3-4

END OR FACE MILLING SOFT MACHINERY STEEL.

Diam. of Mill.	Rev. per Minute.	Speed of Cutter pr. Minute.	Depth of Cut.	Width of Cut.	In Scale of S. M. S.	Under Scale of S. M. S.
1/2"	200	26	1-16"	1-2"	2 1-4"	3 1-2"
	200	26	3-16	1-2	7-8	1 1-4
1	130	27	1-16	1	3 1-4	4 5-8
	130	27	3-8	1	9-16	7-8
1 3/4	31	25	1-16	1 3-4	2 1-4	3 3-8
	31	25	19-32	1 3-4	1-4	3-8

SURFACE MILLING CAST IRON.

Diam. of Mill.	Revs. per Minute.	Speed of Cutter pr. Minute.	Depth of Cut.	Width of Cut.	In Scale of Cast Iron.	Under Scale of Cast Iron.
3	51	40	1-16	1	8	12
	51	40	1-2	1	3 1-4	4 3-4
	51	40	1-16	2	5	8
	51	40	1-2	3	3 1-4	3 1-4
	51	40	1-16	3	7-8	5
	51	40	1-2	4	3 1-8	5
4 1/2	32	38	1-16	4	2	3 1-8
	32	38	1-16	6	2	3 1-8
	32	38	1-16	6	1 1-4	1 3-8
	32	38	1-16	8	2	3 1-8
	32	38	1-4	8	1 3-8	2
	32	38	1-16	10	7-8	1 1-4
	32	38	1-4	12	1 3-8	6
	32	38	1-4	12	1-2	7-8

SURFACE MILLING SOFT MACHINERY STEEL.

Diam. of Mill.	Rev. per Minute.	Speed of Cutter pr. Minute.	Depth of Cut.	Width of Cut.	In Scale of S. M. S.	Under Scale of S. M. S.
3	40	31	1-16	1	6 3-8	9
	40	31	1-2	1	1 1-2	1 6-8
	40	31	1-16	2	1 5-8	2 1-2
	40	31	1-2	3	5-8	1
	40	31	1-16	3	1	1 5-8
	40	31	1-2	4	3 3-8	5
4 1/2	32	38	1-4	4	1 5-16	2
	32	38	1-16	6	1 5-16	2
	32	38	1-4	6	1-2	7-8
	32	38	1-16	8	1	1 5-16
	32	38	1-4	10	11-32	1-2
	32	38	1-16	10	1-2	7-8
	32	38	1-4	12	7-32	11-32
	32	38	1-4	12	7-8	1
	32	38	1-4	12	9-64	7-32

SPEEDS AND FEEDS OF No. 5 PLAIN MILLING MACHINE.

END OR FACE MILLING CAST IRON.

Diam. of Mill.	Rev. per Minute.	Speed of Cutter pr. Minute.	Depth of Cut.	Width of Cut.	Feed per Minute — In Scale of Cast Iron.	Under Scale of Cast Iron.
1/2"	382	50'	1-16"	1-2"		35"
1	382	50	1-8	1-2	7	11
1 3/4	191	50	1-16	1	30	40
	191	50	1-2	1	3	5 1-2
5	109	50	1-8	1 3-4	3 5-8	23
	109	50	3-4	1 3-4	2 5-9	1-8
16	42	55	1-4	5	7-8	4 1-8
	10	45	1-4	16		1

FACE MILLING SOFT MACHINERY STEEL.

Diam. of Mill.	Rev. per Minute.	Speed of Cutter pr. Minute.	Depth of Cut.	Width of Cut.	Feed per Minute — In Scale of S.M.S.	Under Scale of S.M.S.
1/2"	267	35'	1-16"	1-2"		
1	267	35	1-4	1-2	3	4 3-4
1 3/4	152	40	1-16	1		
	152	40	1-2	1	2 3-4	4 1-2
	87	40	1-16	1 3-4		1 3-4
	87	40	3-4	1 3-4		

SURFACE MILLING CAST IRON.

Diam. of Mill.	Rev. per Minute.	Speed of Cutter pr. Minute.	Depth of Cut.	Width of Cut.	In Scale of Cast Iron.	Under Scale of Cast Iron.
3	42	34	1-16	1	6 5-8	8 7-8
	42	34	1-2	1	4 1-8	6 1-2
	42	34	1-16	2	6 5-8	8 7-8
	42	34	1-2	2	2 1-2	4
	42	34	1-16	3	6 5-8	8 7-8
	42	34	7-16	3	1 3-8	2 3-16
3 1/2	42	40	6-32	8	4 1-8	6 1-4
	42	4	1-2	3 1-2	2 1-2	3
4 1/2	42	50	1	2	4 1-8	6 1-2
	42	50	1	4	3 1-2	4 1-8
	42	50	1-8	6	4 1-8	2 1-4
	42	50	11-32	12	1 3-8	1-2

6 = KEYSEATING OR GANG MILLS.

4	42	45	1-16	7-8 each.	6 1-8	8 7-8
	42	45	1-2	7-8 each.	1 3-8	1 1-2

SURFACE MILLING SOFT MACHINERY STEEL.

Diam. of Mill.	Rev. per Minute.	Speed of Cutter pr. Minute.	Depth of Cut.	Width of Cut.	In Scale of S.M.S.	Under Scale of S.M.S.
3	38	30	1-16	1	6	8
	38	30	1-2	1	1 1-8	1 7-8
	38	30	1-16	2	2 3-8	3 1-2
	38	30	1-2	2	3-4	1 1-8
	38	30	1-16	3	1 1-8	1 7-8
	38	30	3-8	3	3-4	1 1-8
3 1/2	38	35	1-16	8	1 7-8	3
	38	35	1-8	8	3 4	1 1-8
4 1/2	25	30	1-16	3	4	5
	25	30	1-16	5	2 1-2	4 1-2
	25	30	3-8	5	1 3-4	1-2
4	25	30	1-16	10	1 3-4	2 1-2
	25	30	3-16	10	1-2	7-8

6 = KEYSEATING OR GANG MILLS.

4	28	30	1-16	7-8 each.	4 3-8	5 7-8
	28	30	1-2	7-8 each.	1-2	7-8

face milling.   But very accurate milling may be done and
excellent surfaces obtained by small end mills running at
high speeds.

Pickling   Castings that are to be milled should be free from sand.
Castings
and   They should be well pickled, and in some cases it is an
Forgings.
advantage to have them rattled after being pickled.
Where they are small and are to be finished rapidly it is
also well to have them annealed.

. Forgings should be free from scale.   They can be
pickled in ten minutes in one part sulphuric acid and
twenty-five parts boiling water, and if then rinsed in boiling water they will dry before becoming rusty.

Washers   When the collars sent with our milling arbors are not
and Collars
for Arbors. the right thickness to bring the cutters into the desired
position washers are employed.   The following thicknesses
are convenient — .001″, .002″, .004″, .008″, .016″, and
.032″ — as these give all steps from .001″ to .032″.   The
collars usually sent are cast iron, but for hard usage steel
collars are preferable.

Lead   A lead hammer is frequently used to drive arbors or
Hammer.
collets into the spindle or bottom or seat work in the
vise.

A bar of brass or copper, $\frac{3}{4}$ inch diameter by 5 or 6
inches long, will also be found useful to place against end
mills, or the end of small collets after the mills are in place.
In this way the driving is often more conveniently done
and any hammer may be used.

Selection   The construction of the machine having been carefully
of Work
for Novice. examined, and the novice having made himself familiar
with the various classes of cutters, it is well if he is using
a Universal Machine, to choose some work at the outset
that does not require the use of the spiral head.   Such
work is generally held in the vise when milled.

Setting   The vise can be set with its jaws parallel to the spindle
Vise with
Plain Base. by placing one of the milling arbors in the spindle and
then bringing the jaws up to the arbor.   It can be set at
right angles with the spindle by a square placed against
the arbor and the jaws.

The front of the table of the machine can also be used in setting the vise.

The swivel vise, which is now sent with all of the Universal Milling Machines and Plain Milling Machines with Screw Feed, can be set by making use of the graduations on the base.

To illustrate the various ways in which the machines are used, we have given descriptions and cuts of a number of operations. Many of these cuts are made on the No. 1 Universal Milling Machine, but it is obvious that the other machines may generally be adapted to similar work, and the directions given in connection with the No. 1 Universal Milling Machine can, in most instances, be applied when doing work on the other machines. *Examples of Work Done on Milling Machines.*

Small stock may be conveniently and advantageously cut into short lengths, especially when it is of a section not easily handled in a cutting off machine. The cutter commonly used is the metal slitting saw, shown on page 126. It is held on an arbor as in Fig. 23, and the work is fed as indicated by the arrow on the piece A. The table may be fed by hand or automatically, the stop being used to determine the length of feed, and the work may be placed exactly in the required position by noting the readings on the graduations around the end of the cross feed shaft. When cut in a Milling Machine the pieces are square at both ends and uniform in length. *Cutting off Stock.*

If it is desirable only to square the ends of small pieces and make them uniform in length, it may be done by first squaring one end of all the pieces, and then by bringing the finished end against a piece clamped to the immovable vise jaw to serve as a stop, while the other end is brought against the mill by the cross feed and run past it automatically or by hand, the vise jaws being parallel with the spindle. In squaring up the ends, at first, a face or an end mill, as shown on page 126, or possibly a side milling cutter, Fig. 23, screwed on the end of an arbor may be used. *Squaring the Ends of Small Pieces.*

Fig. 24 shows two side mills or twin mills upon an arbor. They ordinarily revolve in the direction indicated by the arrow, and are used in milling tongues or ribs.

As shown they are held apart by a collar or washer, but a milling cutter may be put between them and used to dress the top of the tongue and five surfaces then be milled at once.

The two sides and bottom of our Milling Machine Vises, page 151, are milled on the No. 5 Plain Milling Machine, with cutters arranged in this way. The length of cut on the No. 2 Vise is 9½ inches, width on the top 5½ inches, and the height of the sides 2¼ inches, making the total width of cut 10 inches. The side mills have 38 teeth and are 8 inches in diameter, the milling cutter has 14 teeth and is 2¾ inches in diameter. The number of revolutions of the spindle is 24 per minute; the feed is 4½ inches per minute. The time required to set and remove the work and make the cut is eight min-utes. The figures are not given as showing remarkable speed, but as indicating what may be easily attained in ordinary work.

By a similar arrangement of cutters small rectangular pieces can be milled with two cuts, the first across the top and sides, the second across the bottom and ends.

Cast iron pieces 6 inches wide and about 8 inches long are milled on the sides and bottom on the No. 24 Plain Milling Machine, by cutters making 24 revolutions per minute, with a feed of 2 inches per minute. The side milling cutters are 5⅝ inches diameter, and the milling cutters 3 inches diameter. The total width of cut is about 8 inches and the depth $\frac{3}{10}$ inch. The top is milled by an inserted tooth mill 6¾ inches diameter, making 22 revolutions per minute, with a feed of 2⅛ inches. Another cut is made on the inside with an inserted tooth mill 4 inches diameter, making 36 revolutions per minute, with a feed of 2¾ inches.

A guaranteed cut in tool steel on the No. 4 Plain Machine, which was surpassed in daily practice was 8

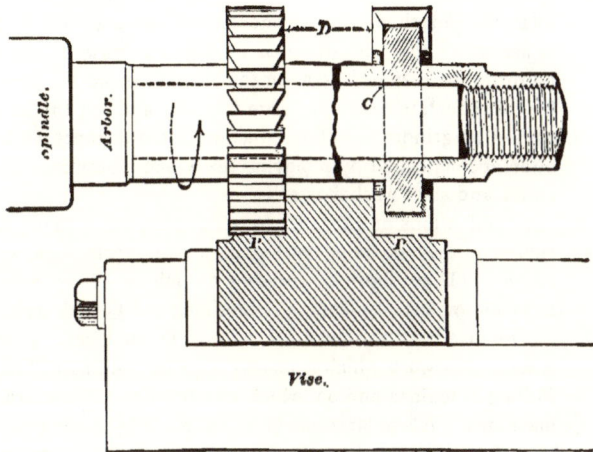

FIG. 24.

MILLING MACHINE VISE.

inches wide, ⅛ inch deep, with ¾ inch feed per minute. And one of the cuts most commonly seen on a No. 23 Plain in our shops is in cast iron 9 inches wide, ⅛ inch deep, with 3 inch feed and 30 feet surface speed per minute.

When side cutting teeth are dull, straddle mills may be interchanged, thus bringing sharp teeth into use, and if it is necessary that the distance D, Fig. 24, must be kept uniform, the distance between the cutters should be capable of reduction to compensate for wear and for the loss caused by grinding. This reduction in the distance D, is often accomplished by removing the washers between the cutter and steel washer as at C.

Combinations or gangs of form cutters work somewhat similar to that indicated by Fig. 24 are shown in Figs. 21 and 25. These figures represent the milling of opposite sides of small steel pieces, 3 inches long and ½ inch wide.

The vise jaws as shown are made to fit the piece. This is a class of work which is usually done on the small Plain Milling Machines and on which one boy can run several machines. A similar class of work, made by a combination of two form and one ordinary milling cutter is shown by Fig. 26.

Fig. 27 shows twin or straddle mills cutting a bolt head or nut. When these mills are not at hand the work can be milled by a milling cutter as in Fig. 28, or by an end mill as in Fig. 29. In either case when nuts are milled they are usually strung on a mandrel.

Fixtures for Small Work. For many kinds of work the fixture shown in. Fig. 30 is convenient. It consists of a square piece of cast iron several inches in length, bored to receive a shaft or spindle to be split at one end or both ends as shown ; or to have a series of holes or flat places made at right angles with, or directly opposite each other. The slot runs the entire length of the fixture and a small screw is inserted at S to hold the work and prevent it from turning while in the shell.

The fixture is held in the vise as in Fig. 30, and the work does not have to be changed in the fixture, but after

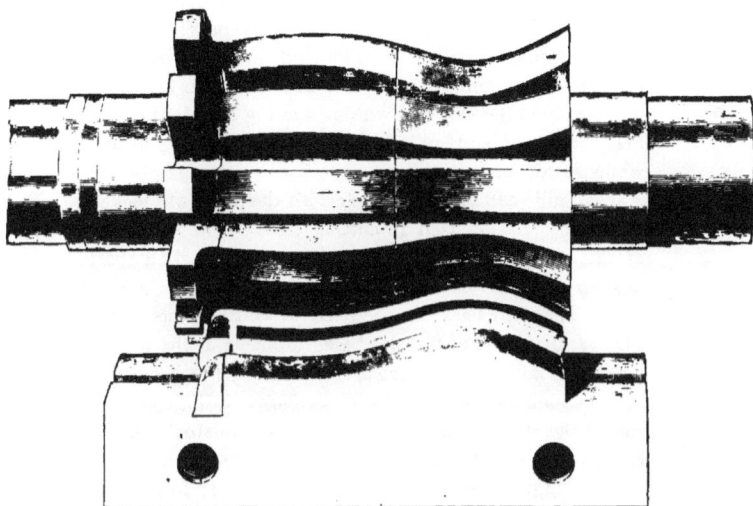

FIG. 25.

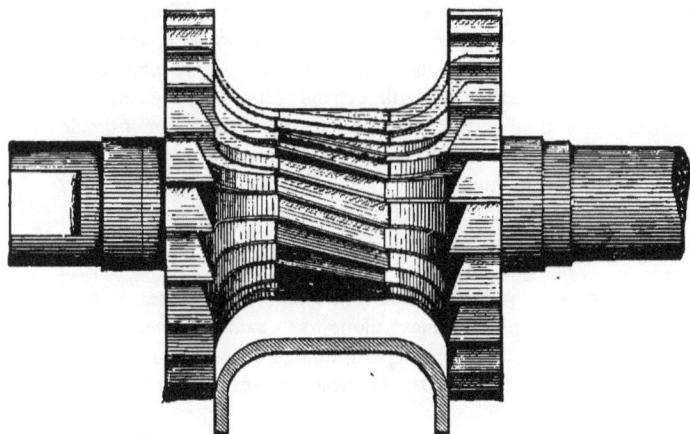

FIG. 26.

each cut is taken, the fixture is given one-quarter of a turn, and each of the four sides of the bolt head or other work brought to the mill.

If the work to be milled is three or six sided, the fixture must be six sided. With two mills as in Fig 27, the fixture would have to be changed in the vise only three times to mill a six sided piece. A fixture can be made without the slot and with a tapering hole for holding the shanks of end mills, twist drills and milling arbors while the tenons are milled.

Cutting Slots.   Two mills can be put together with the teeth interlocked as in Fig. 31, and used in cutting slots. The advantage of this arrangement is that the mills can be blocked apart to keep the width of the cut always the same.

The end mill can also be used in cutting slots, as in Fig. 32, the width of the slot being the same as the diameter of the mill. Shafts that are to be slotted may be held in the chuck on the spiral head, as shown in Fig. 32, one end of the shaft being supported by the foot-stock centre. In this case it is better to feed the work back and forth the full length, than to drill holes at A and B; for, unless unusual care is taken in drilling, they are likely to be out of line. By use of the cross feed the work may be moved in until the seat is cut to the desired depth. For cutting slot some prefer the Cotter mill, page 135. Each of these mills when dropped directly into the work, leaves a teat in the centre, but when the work is fed along to make the slot, the Cotter mill will cut the teat off more easily than the end mill. Neither a Cotter mill nor an end mill should be used in making slots when they can be made with a regular milling cutter, page 126 for milling cutters cut faster and produce slots more uniform in size than either end or Cotter mills.

The End Mill with centre cut, page 126, is useful where it is desired to cut into the work with the end of the mill, and then move along as in cams, grooves, etc., as the teeth have a cutting edge, and are sharp on the inside, and thus cut a path out from the first entering point.

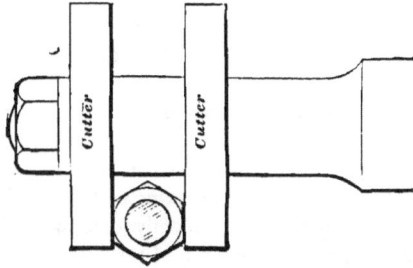

FIG. 27.

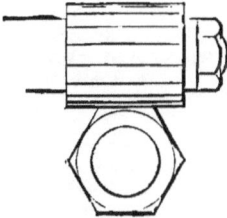

FIG. 28.

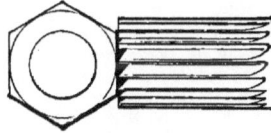

FIG. 29.

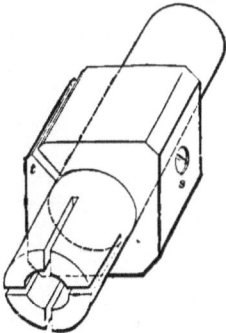

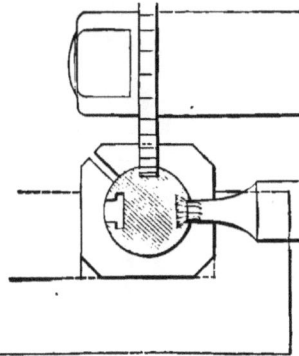

FIG. 30.

They are also useful in taking heavy cuts, especially in cast iron, by reason of the coarse teeth.

When, as in Fig. 32, the centre of the index spindle must be at the same height as the cone spindle, the top of the knee is brought to the line on the frame marked *centre*.

Should it be desirable to go either above or below the centre of the spindle a given distance, the graduated collar on the lower screw shafts may be turned and set at zero ; then by use of the following table of decimal equivalents the distance may be readily determined in eighths, sixteenths, etc., as every turn of the screw shaft will raise or lower the knee one tenth or one hundred one thousandths of an inch.

T SLOT CUTTER.

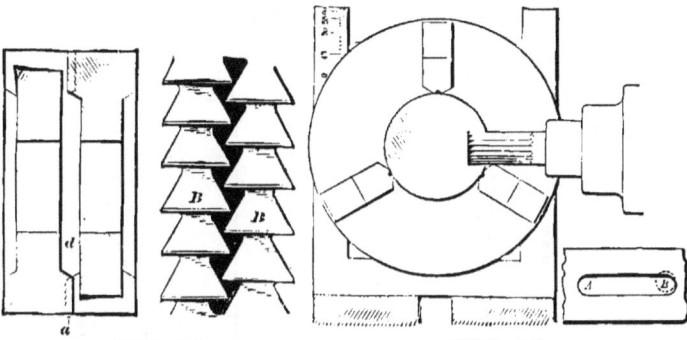

FIG. 31. FIG. 32.

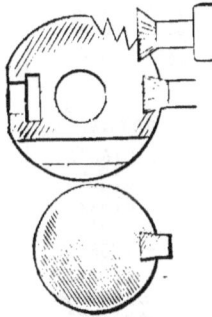

FIG. 33.

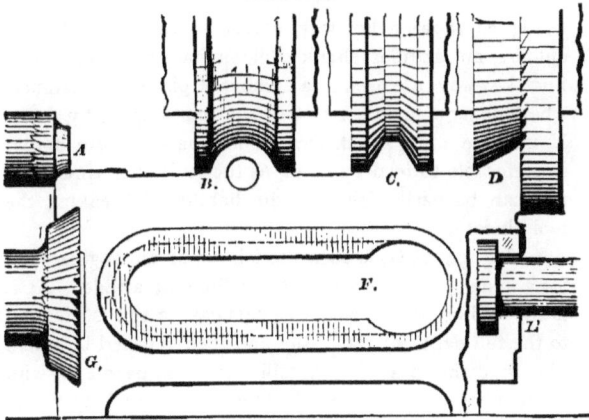

FIG. 34.

TABLE OF DECIMAL EQUIVALENTS.

8ths.		
$\frac{1}{8}=.125$	$\frac{7}{32}=.21875$	$\frac{17}{64}=.265625$
$\frac{1}{4}=.25$	$\frac{9}{32}=.28125$	$\frac{19}{64}=.296875$
$\frac{3}{8}=.375$	$\frac{11}{32}=.34375$	$\frac{21}{64}=.328125$
$\frac{1}{2}=.50$	$\frac{13}{32}=.40625$	$\frac{23}{64}=.359375$
$\frac{5}{8}=.625$	$\frac{15}{32}=.46875$	$\frac{25}{64}=.390625$
$\frac{3}{4}=.75$	$\frac{17}{32}=.53125$	$\frac{27}{64}=.421875$
$\frac{7}{8}=.875$	$\frac{19}{32}=.59375$	$\frac{29}{64}=.453125$
	$\frac{21}{32}=.65625$	$\frac{31}{64}=.484375$
16ths.	$\frac{23}{32}=.71875$	$\frac{33}{64}=.515625$
$\frac{1}{16}=.0625$	$\frac{25}{32}=.78125$	$\frac{35}{64}=.546875$
$\frac{3}{16}=.1875$	$\frac{27}{32}=.84375$	$\frac{37}{64}=.578125$
$\frac{5}{16}=.3125$	$\frac{29}{32}=.90625$	$\frac{39}{64}=.609375$
$\frac{7}{16}=.4375$	$\frac{31}{32}=.96875$	$\frac{41}{64}=.640625$
$\frac{9}{16}=.5625$		$\frac{43}{64}=.671875$
$\frac{11}{16}=.6875$		$\frac{45}{64}=.703125$
$\frac{13}{16}=.8125$	**64ths.**	$\frac{47}{64}=.734375$
$\frac{15}{16}=.9375$	$\frac{1}{64}=.015625$	$\frac{49}{64}=.765625$
	$\frac{3}{64}=.046875$	$\frac{51}{64}=.796875$
32nds.	$\frac{5}{64}=.078125$	$\frac{53}{64}=.828125$
$\frac{1}{32}=.03125$	$\frac{7}{64}=.109375$	$\frac{55}{64}=.859375$
$\frac{3}{32}=.09375$	$\frac{9}{64}=.140625$	$\frac{57}{64}=.890625$
$\frac{5}{32}=.15625$	$\frac{11}{64}=.171875$	$\frac{59}{64}=.921875$
	$\frac{13}{64}=.203125$	$\frac{61}{64}=.953125$
	$\frac{15}{64}=.234375$	$\frac{63}{64}=.984375$

Fig. 33 shows a key-way, widest at the bottom, that is cut by an angular shank mill similar to an end mill. Such key-ways are used in a number of places, for instance, where change gears are to be kept from slipping by a key. If made in a single cut, and the mill be a delicate one, it is preferable to feed by hand, as the resistance and progress can be easily felt, and the liability of breaking the tool obviated.

In Fig. 34, at G, is seen the milling of an angular slot for a slide. At D is shown the milling of a slide. At E is the milling of a T slot. In this case a slot is first cut to the full depth of the T slot and then finished by a mill like that shown at E. A mill shown on page 156, with every other tooth shortened at one end, lasts longer and cuts faster in milling T slots than an ordinary end mill.

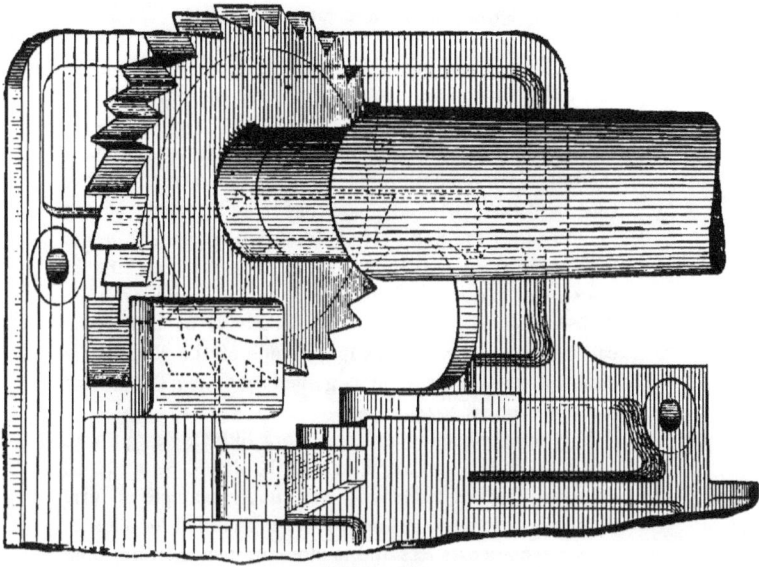

FIG. 35.

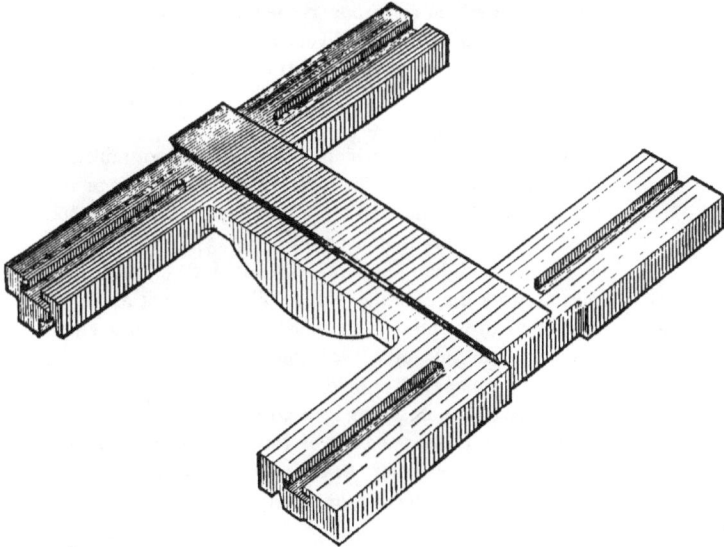

FIG. 36.

Some T slots are enclosed like F, Fig. 34, and to enter this a hole is sometimes made at the back through the body of the piece.

In milling cored T slots in cast iron, it is better to have the mill small enough to cut only one side of the slot at a time, and the side of the slot that is being milled should move against the mill. If the mill is large enough to cut both sides at once, then the teeth strike the scale first in the side of the slot that runs with the mill and are soon dulled. A full size mill is also likely to be broken by catching upon high places in the side of the slot. It is doubtful whether anything is gained by coring T slots that are to be milled.

It is usually economical to do work like C. Fig. 34, with a formed cutter, and anything like B, should certainly be done with such a cutter. A very simple example of form milling is seen at A, where a corner is rounded for a finish. This method costs less and is more perfect than finishing corners with a file, and for such work a mill of large diameter used on an arbor is better than a shank mill, unless the shank mill is run very fast.

Fig. 35 shows a method of cutting a slot or groove with a milling cutter. The piece shown is cast iron and the cuts are to a close limit.

Special Operations.   A lathe carriage in which T slots have been milled is shown in Figure 36. The bearings of this carriage and the ways of the lathe bed are also milled, and the method of doing this work is indicated by Fig. 37.

The teeth of hair clippers shown in Fig. 38, are examples of work done on the No. 12 Plain Milling Machine.

In Fig. 39 is shown a method of milling a surface and squaring one side of a projection on the surface. The vise jaws are special to hold the work in the best manner and to save time in setting it in position.

In Fig. 40 is seen the use of formed cutters in milling rack teeth. The cutter shown is made in three parts and each part cuts six teeth in the rack.

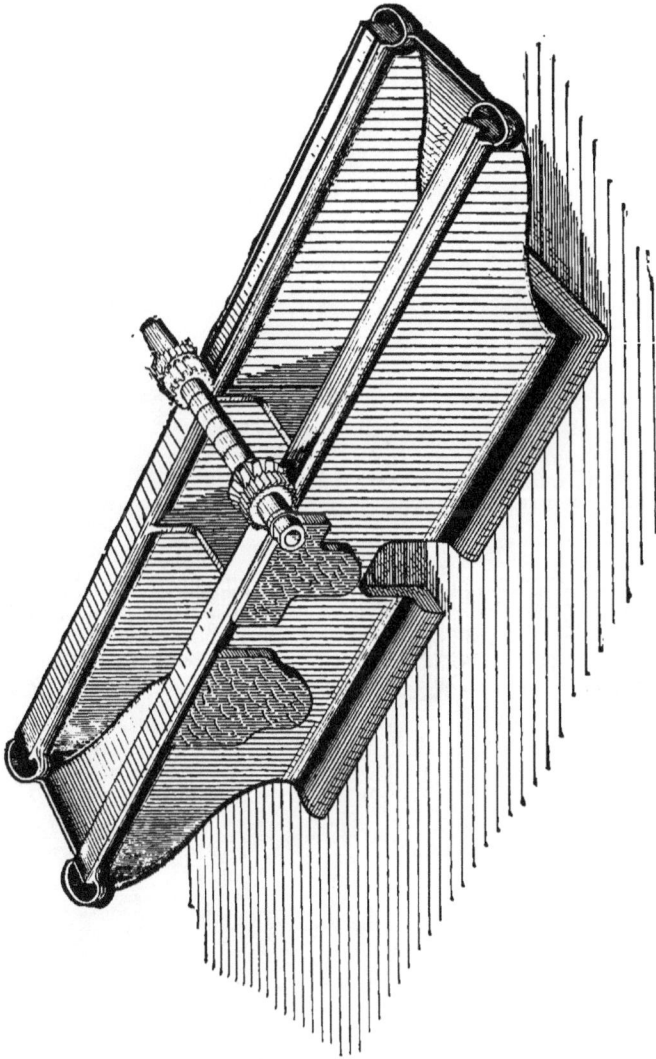

FIG. 37.

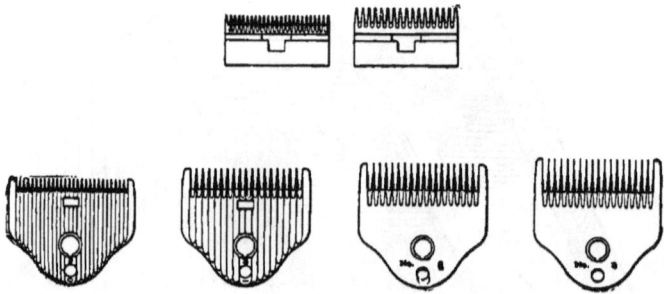

FIG. 38.

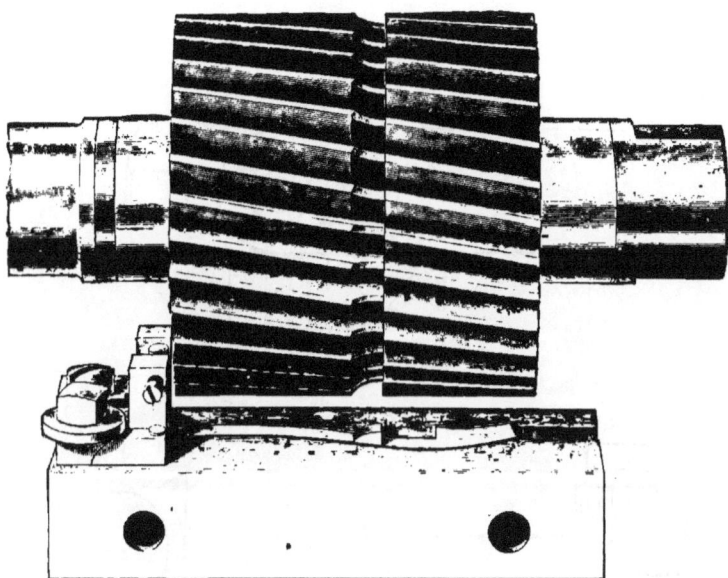

FIG. 39.

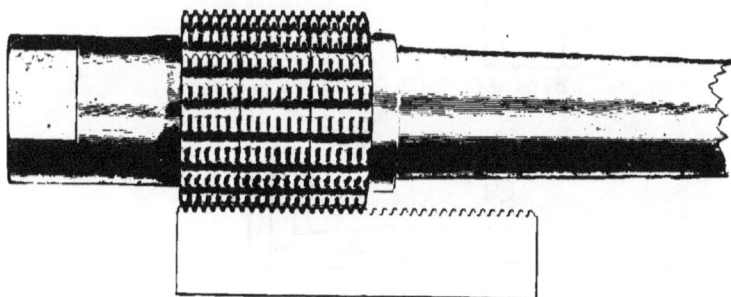

FIG. 40.

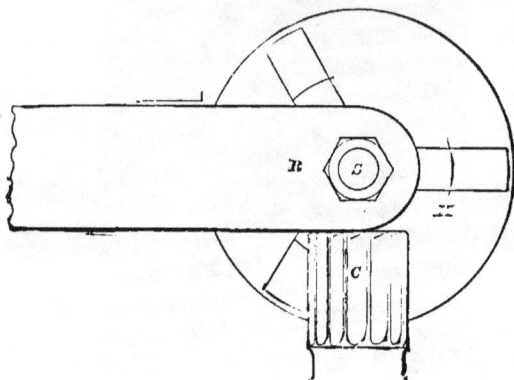

FIG. 41.

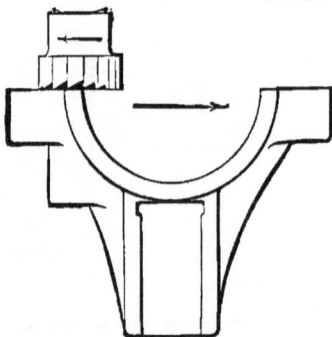

FIG. 42.

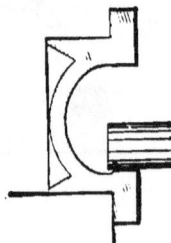

FIG. 43.

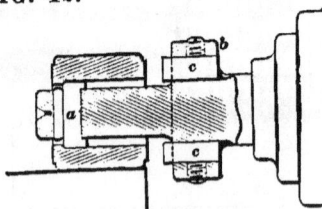

FIG. 44.

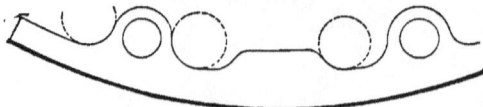

FIG. 45.

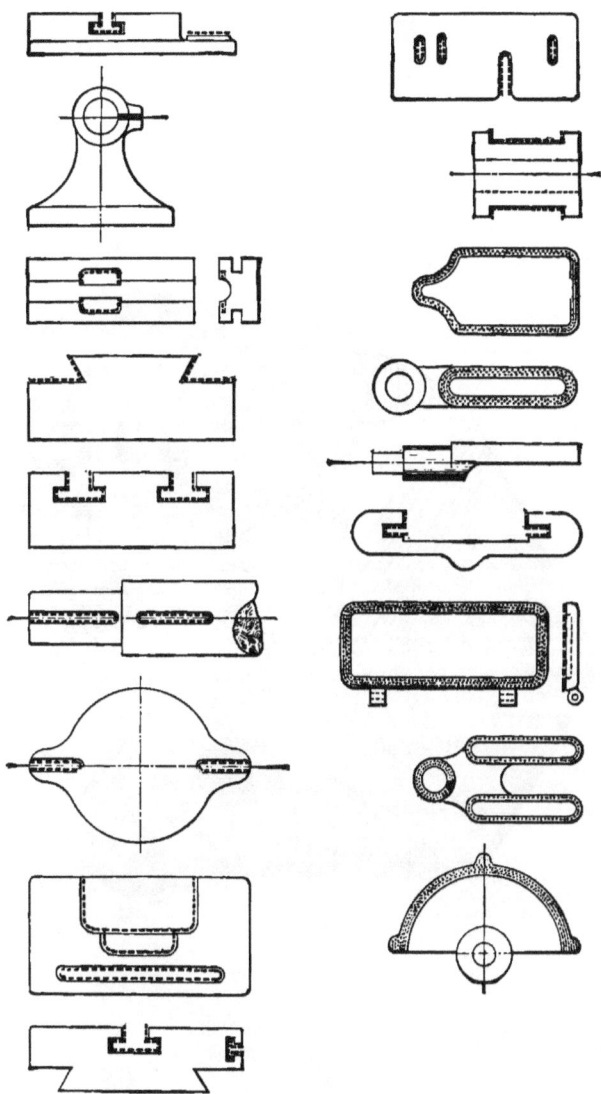

FIG. 46.

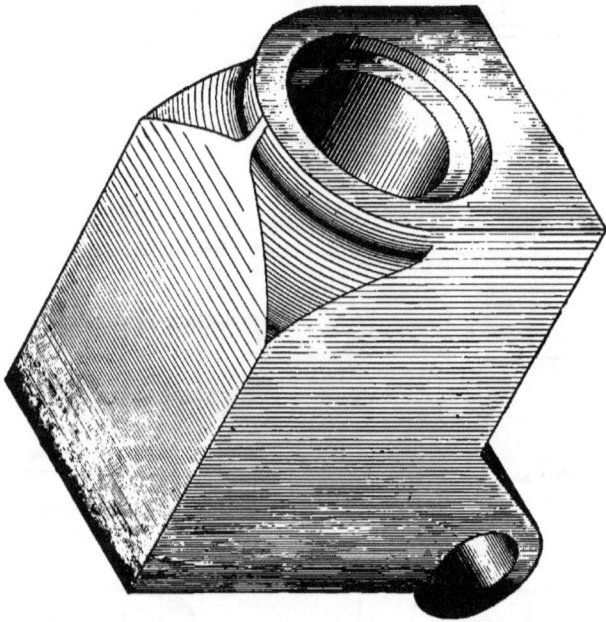

FIG. 47.

When a few pieces are to have round ends they may be milled as in Fig. 41, the piece R being rotated about S against the mill C.

Figs. 42 and 43 are representative operations that need no description.

Fig. 44 shows how a cylinder may be bored and one end squared in a milling machine, the action of the parts of a Plain Milling Machine being quite similar to that of a boring machine.

Fig. 45 shows a method of milling between the arms of hand wheels.

Fig. 46 shows samples of work done on the Vertical Spindle Milling Machine. Many of these cuts may also be made, though less conveniently, on machines with horizontal spindles, and all may be made on a horizontal machine with a Vertical Spindle Milling Attachment.

As an illustration of the number of operations and the variety of work which can profitably be done on the milling machine in the manufacture of machine tools, we give an approximate list of the milling operations performed in our works in the manufacture of the spiral head of the No. 1 Universal Milling Machine. Those marked with a star require the use of an Index Head, but by far the larger part of the work can be done without an Index Head upon our various Universal or Plain Milling Machines, and are examples of what may be practically termed plain milling.

Milling Cuts made in Manufacturing the No. 1 Universal Milling Machine.

LIST OF THE MILLING OPERATIONS IN THE MANUFACTURE OF THE NO. 1 UNIVERSAL MILLING MACHINE SPIRAL HEAD AND FOOT-STOCK.

SPIRAL HEAD, FIG. 47.

NO. OF CUTS.

*Roughing both sides and finishing top............................. 3
Milling bottom and angle as shown in Fig. 48......................... 1
Milling back end.. 1
Milling around lug for worm 1
*Milling front end.. 1
*Bevel cut clearance for mitre gears................................ 1

SPIRAL BOX, FIG. 51.

Milling inside bottom.. 1
Milling recess for nuts on bottom................................... 2
Milling back side... 1
*Milling around circle.. 1
*Milling around circle on top....................................... 2
Milling circular bolt slots 2
Milling for tongue in bottom 1
Milling complete sector. 3

INDEX CRANK J, FIG. 3.

Milling both sides... 2
Milling both edges with a form cutter............................... 2
Milling slot as shown in Fig. 50.................................... 1

WORM SHAFT O, FIG. 3.

*Straddle cut for index crank 1
*Cut for worm key, Fig. 49.. 1

SPIRAL SHELL.

*Milling dovetail key-way similar to Fig. 33........................ 1

FACE PLATE.

Gang cut ends and slot like Fig. 52................................. 2

CENTRE.

Milling spot for set screw.. 1

MITRE GEAR COVER.

Milling complete.. 6

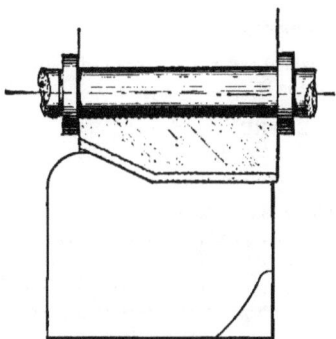

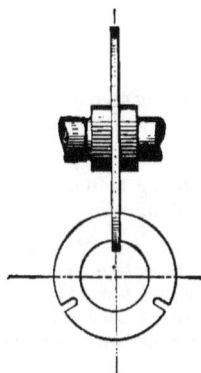

FIG. 48.

FIG. 49.

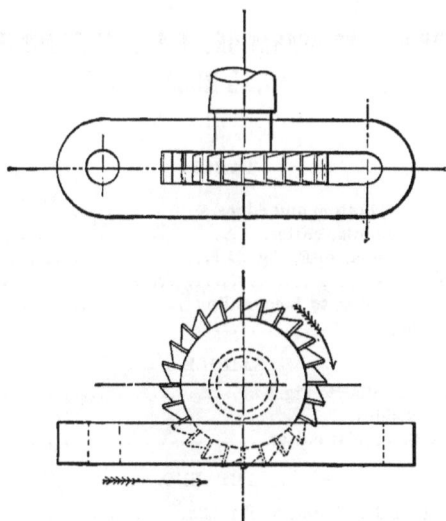

FIG. 50.

MITRE GEAR BLOCK.

Milling two gang cuts, like Fig. 53................................. 2

SHEET STEELS.

Milling around the edges of worm wheel cover and mitre gear cover,
shown in black lines in Figs. 3 and 4...........................11

BACK CENTRE.

Milling bottom......................................:................. 1
Milling sides.. 2

BACK CENTRE STAND.

Milling face 1
Milling top... 1
Milling back side.... 1
Milling bottom similar to Fig. 24.................................... 1

CENTRE REST.

Sawing off part of base... 1
Milling bottom for tongue, similar to Fig. 24........................ 1

LIST OF THE MILLING OPERATIONS IN THE MANUFACTURE OF THE NO. 2 UNIVERSAL MILLING MACHINE.

TABLE.

NO. OF CUTS.

Milling top, sides, bottom and edges................................. 2
Milling Vs with angular cutter...................................... 2
. Milling T slots similar to E, Fig. 34.............................: 2
Milling oil grooves................................. 1
Milling one end similar to face milling............................. 1
Milling table dog slot.... 2

SADDLE.

Milling bottom similar to Fig. 54.................................... 1
Milling top and sides.... 1
Milling Vs with angular cutter....... 1

CLAMP BED.

Milling top and sides similar to Fig. 53............................. 1
Milling bottom similar to Fig. 54.................................. 1
Milling T slots... 2

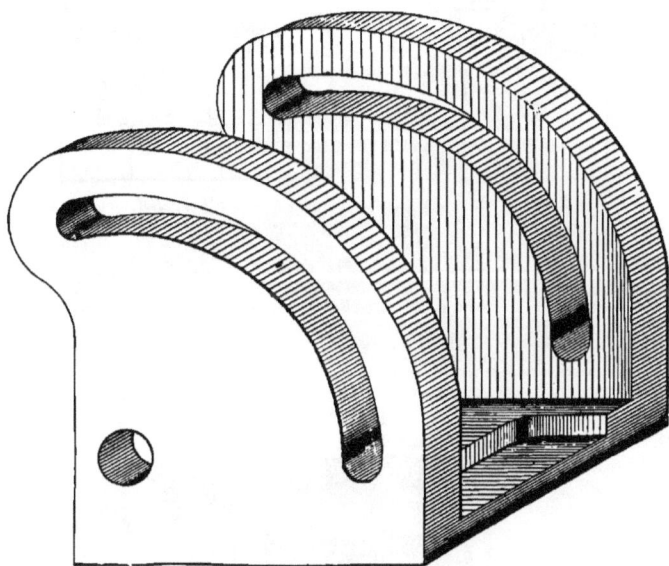

FIG. 51.

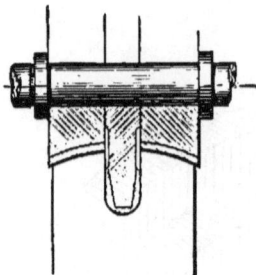

FIG. 52.

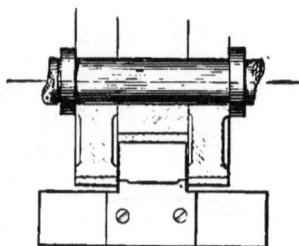

FIG. 53.

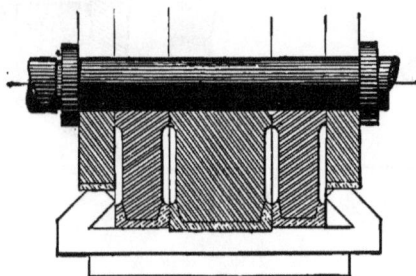

FIG. 54.

KNEE.

TABLE STOP.

TABLE STOP GIBS.

FEED TRIP SLIDE.

FEED ROCKER.

FEED REVERSING LEVER.

TABLE SCREW BEARING.

FEED CLUTCH FORK.

TABLE SCREW.

CONE GEAR GUARD (Centre).

SPINDLE AND CONE GEAR GUARDS (Front and Rear.)

CLUTCH SLEEVE.

KNEE SCREW.

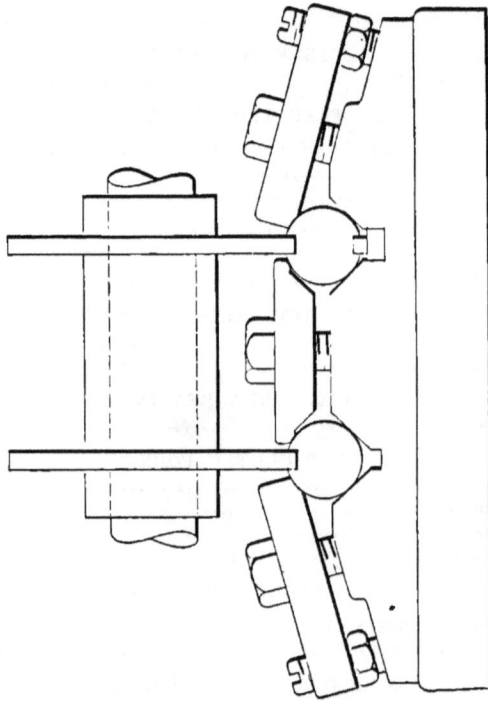

MILLING KEY-WAYS.

KNEE SHAFT.

NO. OF CUTS.

Milling key-way.. 1

CROSS FEED SCREW.

Milling key-way.. 1
Milling square on end.. 4

SPINDLE.

Milling key-ways... 3

STOP ROD.

Milling key-way.. 1

VERTICAL FEED SHAFT.

Milling key-way.. 1

KNEE STOPS.

Milling clamp slot................................... 1

TABLE DOG.

Milling front side... 1
Milling back side.. 1
Milling form on top.. 1

CONE.

Milling dog slot.. 1

CONE DOG BOLT.

Milling flats.. 2

ARM CENTRE.

Milling key-way.. 1
Milling knob slot.. 1

LOWER FEED SHAFT SLEEVE.

Milling key-way.. 1

INTERNAL FEED PULLEY SLEEVE.

Milling key-way.. 1

SADDLE SWIVEL.

Milling key-way.. 1
Milling slot... 1

CLUTCH GEAR BEARING CAPS R. AND L.

Milling seat... 2

SPINDLE FRONT BOX.

Milling slot... 1

SPINDLE REAR BOX.

Milling slot.. 1

USE OF MILLING MACHINES.

INDEXING.

Use of Indexing Head to Divide Periphery of Work into Equal Parts.   The first office of the indexing head stock or spiral head, as mentioned in the description of the machines is to divide the periphery of a piece of work into a number of equal parts, and in connection with the foot stock, it also enables the milling machine to be used for work sometimes done on planer centres and on gear cutting machines.

Direct Indexing.   As the index spindle may be revolved by the crank, and as forty turns of the crank make one revolution of the spindle, to find how many turns of the crank are necessary for a certain division of the work, or what is the same thing, for a certain division of a revolution of the spindle, forty is divided by the number of the divisions which are desired.   The rule then, may be said to be, divide forty by the number of divisions to be made and the quotient will be the number of turns, or the part of a turn, of the crank, which will give each desired division. Applying this rule—to make forty divisions the crank would be turned completely around once to obtain each division, or to obtain twenty divisions it would be turned twice. When, to obtain the necessary divisions, the crank has to

Index Plate.   be turned only a part of the way around, an index plate is used.   For example: If the work is to be divided into eighty divisions the crank must be turned one-half way around, and an index plate with an even number of holes in one of the circles would be selected, it being necessary only to have two holes opposite to each other in the plate.   If the work is to be divided into three divisions an index plate should be selected which has a circle with a number of holes that can be divided by three, as fifteen

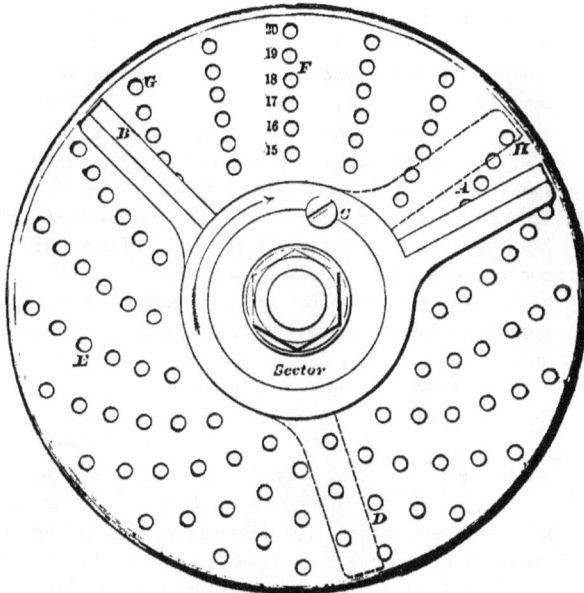

FIG. 55.

or eighteen in Fig. 55, the numbers on the index plates indicating the number of holes in the various circles.

Sector.   The sector is of service in obviating the necessity of counting the holes at each partial rotation or turn of the crank, and to illustrate its use it may be supposed that it is desired to divide the work into 144 divisions.   Dividing 40 by 144 the result, $\frac{5}{18}$, shows that the index crank must be moved $\frac{5}{18}$ of a turn to obtain each of the 144 divisions. An index plate with a circle containing eighteen holes or a multiple of eighteen, is selected, and the sector is set to measure off five spaces or the corresponding multiple, ten spaces for example, in a circle with thirty-six holes. When the sector is set it is held in place by the screw C, Fig. 55.   In setting the sector it should be remembered that there must always be one more hole between the arms than there are spaces to be counted or measured off. Starting with the crank pin in the hole B, for example, a cut would be made in the work and then the crank would be turned and the pin brought into the hole at A and a second cut made in the work.   The sector would then be moved to the position shown at dotted line, that is the arm would be brought against the pin in the crank.   When the third cut is to be made the crank would be turned and the pin brought to the hole D.   The next step would bring the pin to the hole E, and so on.   When the operation has been repeated 144 times one revolution of the spindle will have been made.   The required number of turns of the crank for a large number of divisions may be readily ascertained from   the   accompanying index tables.

Effect of Change in Angle of Elevation of Spindle.   If the angle of elevation of the spiral head spindle is changed during the progress of the work, the work must be rotated slightly to bring it back to the proper position, as when the spindle is elevated or depressed, the worm wheel is rotated about the worm, and the effect is the same as if the worm were turned in the opposite direction.

Examples of Work. Milling Bolt Heads.   A simple use of the index centres is that of milling bolt heads, nuts, etc.   Those shown in Figs. 27, 28 and 29, for

INDEX TABLE.

Number of Divisions.	Number of Holes in the Index Circle.	Number of Turns of the Crank.	Number of Divisions.	Number of Holes in the Index Circle.	Number of Turns of the Crank.
2	ANY	20	35	49	$1\frac{5}{35}$
3	39	$13\frac{13}{39}$	36	27	$1\frac{3}{27}$
4	ANY	10	37	37	$1\frac{3}{37}$
5	"	8	38	19	$1\frac{1}{19}$
6	39	$6\frac{26}{39}$	39	39	$1\frac{1}{39}$
7	49	$5\frac{35}{49}$	40	ANY	1
8	ANY	5	41	41	$\frac{40}{41}$
9	27	$4\frac{12}{27}$	42	21	$\frac{20}{21}$
10	ANY	4	43	43	$\frac{40}{43}$
11	33	$3\frac{21}{33}$	44	33	$\frac{30}{33}$
12	39	$3\frac{13}{39}$	45	27	$\frac{24}{27}$
13	39	$3\frac{3}{39}$	46	23	$\frac{20}{23}$
14	49	$2\frac{28}{49}$	47	47	$\frac{40}{47}$
15	39	$2\frac{26}{39}$	48	18	$\frac{15}{18}$
16	20	$2\frac{10}{20}$	49	49	$\frac{40}{49}$
17	17	$2\frac{6}{17}$	50	20	$\frac{16}{20}$
18	27	$2\frac{6}{27}$	52	39	$\frac{30}{39}$
19	19	$2\frac{2}{19}$	54	27	$\frac{20}{27}$
20	ANY	2	55	33	$\frac{24}{33}$
21	21	$1\frac{19}{21}$	56	49	$\frac{35}{49}$
22	33	$1\frac{27}{33}$	58	29	$\frac{20}{29}$
23	23	$1\frac{17}{23}$	60	39	$\frac{26}{39}$
24	39	$1\frac{26}{39}$	62	31	$\frac{20}{31}$
25	20	$1\frac{12}{20}$	64	16	$\frac{10}{16}$
26	39	$1\frac{21}{39}$	65	39	$\frac{24}{39}$
27	27	$1\frac{13}{27}$	66	33	$\frac{20}{33}$
28	49	$1\frac{21}{49}$	68	17	$\frac{10}{17}$
29	29	$1\frac{11}{29}$	70	49	$\frac{28}{49}$
30	39	$1\frac{13}{39}$	72	27	$\frac{15}{27}$
31	31	$1\frac{9}{31}$	74	37	$\frac{20}{37}$
32	20	$1\frac{5}{20}$	75	15	$\frac{8}{15}$
33	33	$1\frac{7}{33}$	76	19	$\frac{10}{19}$
34	17	$1\frac{3}{17}$	78	39	$\frac{20}{39}$

INDEX TABLE.

Number of Divisions.	Number of Holes in the Index Circle.	Number of Turns of the Crank.	Number of Divisions.	Number of Holes in the Index Circle.	Number of Turns of the Crank.
80	20	$\frac{10}{20}$	164	41	$\frac{10}{41}$
82	41	$\frac{20}{41}$	165	33	$\frac{8}{33}$
84	21	$\frac{10}{21}$	168	21	$\frac{5}{21}$
85	17	$\frac{8}{17}$	170	17	$\frac{4}{17}$
86	43	$\frac{20}{43}$	172	43	$\frac{10}{43}$
88	33	$\frac{15}{33}$	180	27	$\frac{6}{27}$
90	27	$\frac{12}{27}$	184	23	$\frac{5}{23}$
92	23	$\frac{10}{23}$	185	37	$\frac{8}{37}$
94	47	$\frac{20}{47}$	188	47	$\frac{10}{47}$
95	19	$\frac{8}{19}$	190	19	$\frac{4}{19}$
98	49	$\frac{20}{49}$	195	39	$\frac{8}{39}$
100	20	$\frac{8}{20}$	196	49	$\frac{10}{49}$
104	39	$\frac{15}{39}$	200	20	$\frac{4}{20}$
105	21	$\frac{8}{21}$	205	41	$\frac{8}{41}$
108	27	$\frac{10}{27}$	210	21	$\frac{4}{21}$
110	33	$\frac{12}{33}$	215	43	$\frac{8}{43}$
-115	23	$\frac{8}{23}$	216	27	$\frac{5}{27}$
116	29	$\frac{10}{29}$	220	33	$\frac{6}{33}$
120	39	$\frac{13}{39}$	230	23	$\frac{4}{23}$
124	31	$\frac{10}{31}$	232	29	$\frac{5}{29}$
128	16	$\frac{5}{16}$	235	47	$\frac{8}{47}$
130	39	$\frac{12}{39}$	240	18	$\frac{3}{18}$
132	33	$\frac{10}{33}$	245	49	$\frac{8}{49}$
135	27	$\frac{8}{27}$	248	31	$\frac{5}{31}$
136	17	$\frac{5}{17}$	260	39	$\frac{6}{39}$
140	49	$\frac{14}{49}$	264	33	$\frac{5}{33}$
144	18	$\frac{5}{18}$	270	27	$\frac{4}{27}$
145	29	$\frac{8}{29}$	280	49	$\frac{7}{49}$
148	37	$\frac{10}{37}$	290	29	$\frac{4}{29}$
150	15	$\frac{4}{15}$	296	37	$\frac{5}{37}$
152	19	$\frac{5}{19}$	300	15	$\frac{2}{15}$
155	31	$\frac{8}{31}$	310	31	$\frac{4}{31}$
156	39	$\frac{10}{39}$	312	39	$\frac{5}{39}$
160	20	$\frac{5}{20}$	360	18	$\frac{2}{18}$

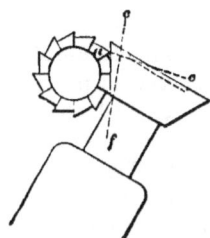

FIG. 56.

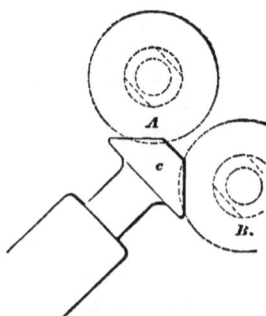

FIG. 57.

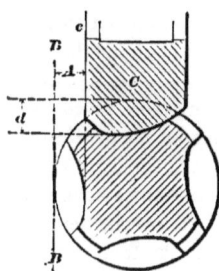

FIG. 58.

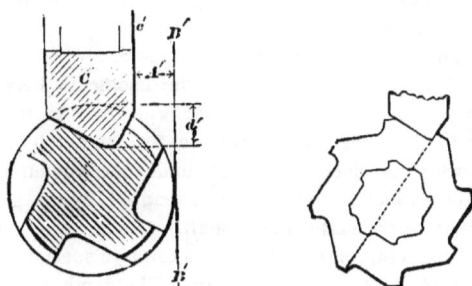

FIG. 59.

instance, could be milled by being held on the spiral head instead of the fixture previously described.

Dividing Cutters, etc.　Among work commonly divided on the index centres are saws, page 126; screw slotting cutters, page 126; and a large variety of straight tooth mills, as shown on page 126. Cutting the teeth of ratchets represents a similar use of the centres. In milling teeth of angular cutters the index spindle can be set as shown in Fig. 56, and when the spindle and cutter are in this position the knee of the machine is raised by the vertical feed. Fig. 57 shows that the same cutter may be used for making a right or left hand mill, in one case the feed being vertical and the other horizontal.

Fluting Taps and Reamers.　The fluting or grooving of taps and reamers is another kind of work in which index centres are used. A table showing the tap and reamer cutters to be used in grooving reamers is given in our catalogue, different cutters being used for a certain size reamer than for the same size tap. The accompanying table gives the figures for setting taps sidewise and also for the depths of groove. These figures will be found convenient, but are not given to be followed absolutely, for mechanics have different opinions as to the best shape and the proper depth of groove in taps.

The columns in the table as marked refer to the special tap cutters. the shape of which are shown in Fig. 58, and tap and reamer cutters Fig. 59. One class of our twist drill cutters have a round edge, and can also be used in grooving taps.

The top of the table of the machine when doing this work is moved sidewise along the distance A, between B B and e, which is .125 of an inch, as given in the column under A. The work is next set for the depth of the cut d, which is given in the table under the column d, and is .2 of an inch. This depth is read off by the graduated collar on the vertical feed shaft.

The depth of groove cut in a tap by a special tap cutter is .2 of the diameter of the tap. The depth of groove cut by a tap and reamer cutter is .23 of the diameter of the tap.

Table for Setting when Cutting with Special Tap Cutters, or with Tap and Reamer Cutters.

Diameter of Tap.	No. of Cutter.	Value of A, *d* and A′, *d*′				Diameter of Cutter.
		A=	*d*=	A′=	*d*′=	
1/8	1	.025	.025	.000	.028	1 3/4
5/32	2	.003	.031	*	.035	"
3/16	"	.018	.037	.000	.043	"
7/32	"	.034	.043	.015	.050	"
1/4	"	.050	.050	.031	.057	"
9/32	3	.028	.056	.015	.065	1 7/8
5/16	"	.043	.062	.031	.072	"
3/8	"	.075	.075	.062	.086	"
7/16	4	.031	.087	.031	.100	2
1/2	"	.062	.100	.062	.115	"
9/16	"	.093	.112	.093	.129	"
5/8	"	.125	.125	.125	.144	"
11/16	5	.081	.137	.093	.158	2 1/8
3/4	"	.112	.150	.125	.172	"
13/16	"	.143	.162	.156	.187	"
7/8	"	.175	.175	.187	.200	"
15/16	6	.093	.187	.125	.215	2 1/4
1	"	.125	.200	.156	.230	"
1 1/16	"	.156	.212	.187	.245	"
1 1/8	"	.187	.225	.218	.258	"
1 3/16	"	.218	.237	.250	.273	"
1 1/4	"	.250	.250	.281	.287	"
1 5/16	7	.168	.262	.218	.300	2 3/8
1 3/8	"	.200	.275	.250	.316	"
1 7/16	"	.231	.287	.281	.330	"
1 1/2	"	.262	.300	.312	.345	"
1 9/16	"	.293	.312	.344	.359	"
1 5/8	"	.325	.325	.375	.373	"
1 11/16	8	.243	.337	.312	.388	2 5/8
1 3/4	"	.275	.350	.344	.400	"
1 7/8	"	.337	.375	.406	.430	"
2	"	.400	.400	.468	.460	"

* No. 2 Regular Cutter, 3/16″ thick.

The machine, as a rule, is fed automatically while cutting, and after each cut is taken the index crank is turned as previously described.

Screw Machine Tool. Fig. 60 shows a tool used in a screw machine or turret lathe. The three views show the results of the various Milling Machine cuts.

Cutting Gears. In cutting gears care must be taken to have the cutter central with the index centres, and to have the cut the exact depth required. A good way of testing the setting is to cut a groove in a piece on the centres, then shift the piece end for end and try the groove upon the cutter. A good method of holding the gear blanks is on an arbor with a taper shank which fits in the index spindle, the outer end of the arbor being supported by the foot stock centre. Frequently in cutting gears we use a shank arbor with expanding bushing and a nut on the arbor at each end of the bushing, one nut forcing the bushing up on the arbor and holding the gear blank, while the other pushes the bushing off the taper and releases the gear when finished. If the common arbor and dog are used, care should be taken that the dog does not spring the arbor. The screws, Fig. 2, may be used to hold the dog so that there shall be no back lash between the index spindle and the work. The depth of the cut can be gauged from the outside of the blank, or, if desired, marked on the side by a gear tooth depth gauge. In cutting gears, when the blank has been placed in position it is raised by the elevating screw until it just touches the cutter. The graduated collar on the vertical feed shaft is placed on zero and the blank moved horizontally from the cutter. Then the work is raised the number of thousandths of an inch required for the depth of tooth.

In our catalogue we give the depth of gear teeth of a number of pitches, directions regarding sizing and cutting of gear wheels, formulae for determining the dimensions of small gears by diametral pitch, also directions for selecting involute and epicycloidal gear cutters for any given pitch. The catalogue contains as well a number of

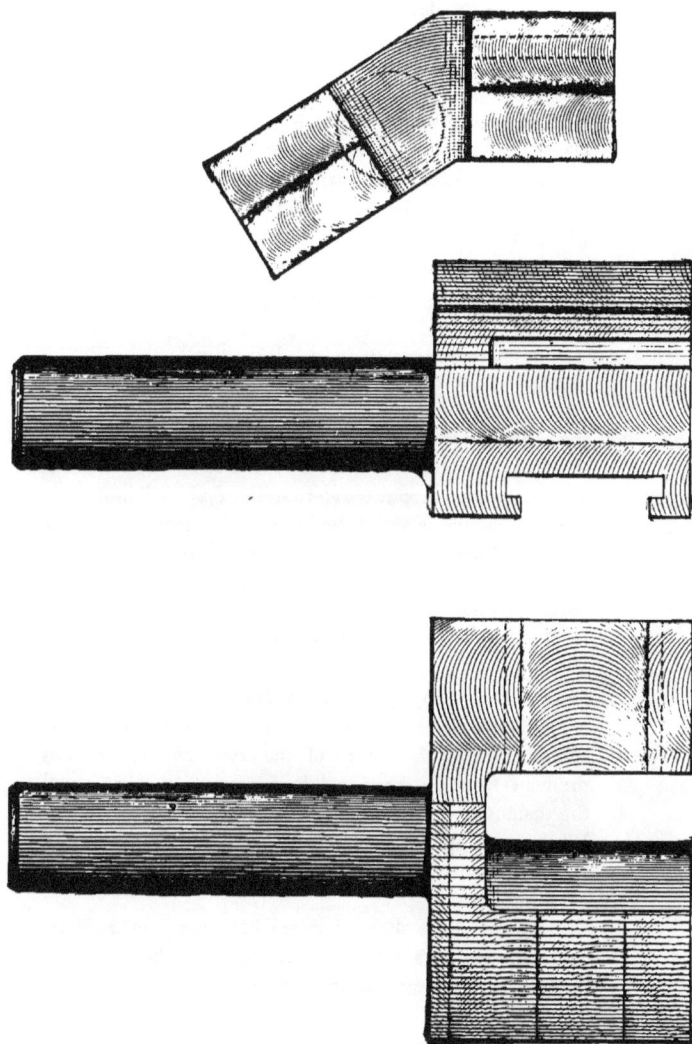

FIG. 60.

plain rules for determining various dimensions of gears. More complete information on this subject is given in our " Practical Treatise on Gearing," and " Formulas in Gearing."

Worm Wheels. For several years we have milled the faces of worm wheels in our Universal Milling Machines, as this costs less and is better than turning them in an engine lathe. The operation will be understood from Fig. 61, one part of which shows a segment of a spur wheel, and the other part S, a segment of a worm wheel blank.

The practice involved in cutting worm wheels is seen in Fig. 62.

The index centres are set central with the cutter on the line C D, as in cutting spur gears.   The arbor holding the worm wheel blank is put on the centres, and by moving the table lengthwise the centre of the face of the worm wheel is set under the centre of the cutter spindle A B. The table stop is put on so that the table will not move, then the saddle is set to the angle of the teeth as seen by the lines E and C D, and the vertical feed is used in cutting the work.   Worm wheels can be hobbed in the Universal Milling Machine.

The wheel can be hobbed to the right depth by setting a rule on the top of the knee and measuring up to the line marked centre on the side of the frame, for when the back of the knee is at this line the index centres are at the same height as the centre of the cone spindle, the measurement from the knee to this line being, in other words, the distance between the centres of the worm and wheel. When hobbing a worm wheel the shaft of which is at right angles with the axis of the worm, the spiral bed should be set at zero.

On page 187 is shown a worm hob that is made in the same manner as a formed cutter, and can be sharpened by grinding without changing its form.

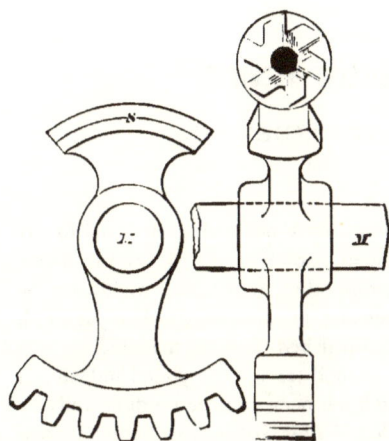
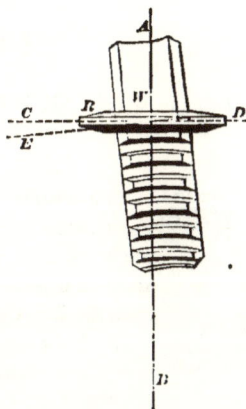

FIG. 61.   FIG. 62.

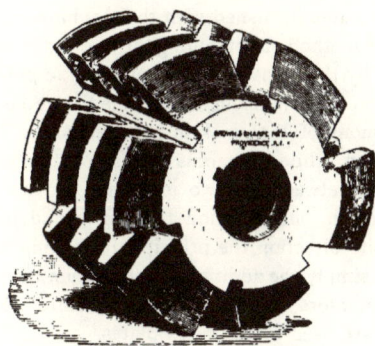

WORM HOB WITH RELIEVED TEETH.

USE OF MILLING MACHINES.

COMPOUND INDEXING.

Compound
Indexing
Defined. This system of indexing is not ordinarily used to any great extent. Frequently, however, it is desired to divide the work into divisions other than those which can be attained by direct indexing, method described on page 176, from the index plates furnished with the machine.

In this case the method termed compound indexing is used ; a method which has been considered quite fully by Messrs. Fred J. Miller, Walter Gribben, G. Schneider and J. V. Hamilton, and explained by them in contributions to the "American Machinist," October 31st, December 12th, 1889, and January 16th, May 22nd, July 17th, August 14th, and October 16th, 1890.

On the following pages we offer tables, for which we are chiefly indebted to the kindness of Mr. Walter Gribben, by the use of which these results can be obtained with the index plates furnished with the machines.

The method of combining the two index settings is indicated by the signs plus and minus; the plus sign showing that the two indexings are added together, or that the movement of the work in both indexings is in the same direction, while the minus sign shows that we take the difference between the two indexings, or that we move one indexing in one direction and the second indexing in the opposite direction, the principle in making the calculations being simply the addition or subtraction of fractions. The denominators of the fractions indicate which circles of holes are used in the index plates.

Direct indexing is obtained by moving the pin straight ahead on a single circle. Tables are given on pages 179 and 180, showing the divisions that can be obtained by direct indexing.

If desired special index plates can be furnished, but in most cases it will be found cheaper and quicker to use the compound indexing method, as described above.

To illustrate the manner of using the machine in com- Manner of
Using the pound indexing, it may be supposed that we desire to divide Machines. the work into 69 parts. Reference to the table, page 191, shows that the work is moved through 21 spaces, or holes in the 23 hole circle and then turned in the opposite direction 11 holes in the 33 hole circle of one of the index plates. The first movement is made in the ordinary manner. The stop or back pin is placed in the 33 hole circle, the index crank pin is pulled out of the 23 hole circle, and the index crank is turned through 21 holes in the desired direction, the holes being measured by the sector. For the second movement, the index crank pin is left in the 23 hole circle, the back pin is pulled back from the plate, and as the minus sign is given in the table, the crank is turned 11 holes in the direction opposite to that of the former movement. In this part of the indexing the index plate and crank turn together, and as there is no sector on the back of the plate, the holes or spaces have to be counted directly in the plate. Had the plus sign been given, as in the indexing to obtain 77 divisions of the work, both movements of the crank would have been in the same direction. Ordinarily the order of the movements is not material and if more convenient for any reason, the back pin could usually be withdrawn first, and the movement described as the second could be made first. In some instances indeed, for example, in dividing the work into 174, 272 or 273 parts, the outer circle is naturally used first.

The correctness of the operations indicated by the table Proof of
Correct- is easily appreciated when the fractions are added or sub- ness of the
Moves. tracted. For example: to divide the work into 69 parts the figures are $\frac{21}{23} - \frac{11}{33} = \frac{21}{23} - \frac{1}{3} = \frac{63}{69} - \frac{23}{69} = \frac{40}{69}$; and this equals one division of the work or the desired result, as there are 40 teeth in the worm wheel, and to obtain 1-69 of a complete revolution it is desired that 40 should be divided by 69.

To obtain 77 divisions the figures are $\frac{0}{21} + \frac{3}{33} = \frac{3}{7} + \frac{1}{11} = \frac{33}{77} + \frac{7}{77} = \frac{40}{77} =$ one division.

This method of calculation (although fractions are involved) is essentially the same as that described on page 176.

Approximate Indexings by Compound Method. The next table gives the movements for divisions of the work (absolutely, or within close limits) from 50 to 250. Where only one indexing is indicated the figures are taken from the tables on pages 179 and 180, and where absolutely correct results are shown by the compound indexing, the figures are taken from the last table. In regard to this table, Mr. Gribben writes as follows :

" The figures in the first column give the desired number of teeth, or the nearest approximations to the desired number of teeth, which approximations are so near in the instance of 51, 53, 57, 61, 67, 71, 73, 79, 81, etc., that the error may be safely neglected.

The figures in the second column give the moves for the index shaft, in opposite directions when connected by a — sign, and in the same direction when connected by a + sign.

The figures in the third column show the error in position of the last space cut if the cross-slide be not moved, and a gear of 1 diametral pitch is being cut. A 1 pitch gear is larger than would be cut on a small machine, but is taken for the reason that from these figures can be computed the the error for finer pitches. This is obtained by dividing the error for 1 pitch by the number of the finer pitch, and the result (if of sufficient magnitude to be taken into account) distributed up around the circumference as best it may.

For instance, if we wish to cut 112 teeth of 6 pitch, the error in the last cut would be about .001 inch, which might be corrected by moving the cross-slide .0001 inch after every 11 cuts. (See page 196.)

Compound Index Movements on the Nos. 1, 2, 3 and 4, Design 1893, Universal Milling Machines. (Absolute and Approximate.)

NO. OF TEETH.	MOVES.	ERROR ON ONE DIAMETRAL PITCH GEAR.	TIMES AROUND.
50.	18/20		
51.0001	8 1/4 − 1 2/18	.0002	11
52.	38/20		
53.0001	6 1/4 − 4/49	.0002	
54.	29/27		
55.	24/33		
56.	34/28		
56.9999	4 12/19 + 7/49	.0003	7
58.	20/28		
58.9997	7 19/40 + 12/18	.0009	11
60.	42/48		
60.9999	3 14/17 + 2/49	.0003	6
62.	39/21		
62.9904	4 13/18 + 1 3/4	.0019	8
64.	10/16		
65.	34/44		
66.	39/46		
67.0002	2 4/17 + 1 4/8	.0005	5
68.	19/40		
69.	24/41 − 11/43		
70.	42/48		
70.9999	3 14/41 − 22/43	.0005	6
72.	19/42		
73.0001	6 14/41 − 4/49	.0002	12
74.	39/42		
75.	1 8/45		
76.	19/40		
77.	9/21 + 3/33		
78.	39/40		
79.0002	2 14/43 + 2/49	.0005	6
80.	19/40		
80.9999	5 4/41 − 9/49	.0003	10
82.	19/41		
83.0003	3 14/43 − 2/49	.0011	8
84.	19/21		
85.	1 8/17		
86.	24/43		
87.	19/43 − 11/44		
88.	14/48		
89.0001	3 14/43 − 4/49	.0005,	8
90.	14/27		

Compound Index Movements on the Nos. 1, 2, 3 and 4, Design 1893, Universal Milling Machines. (Absolute and Approximate.)

NO. OF TEETH.	MOVES.	ERROR ON ONE DIAMETRAL PITCH GEAR.	TIMES AROUND.
91.	$\frac{6}{39} + \frac{14}{33}$		
92.	$\frac{12}{23}$		
93.	$\frac{6}{31} + \frac{14}{33}$		
94.	$\frac{10}{47}$		
95.	$\frac{6}{19}$		
96.	$\frac{6}{18} + \frac{5}{20}$		
96.9999	$4\frac{27}{41} - \frac{6}{19}$	.0004	11
98.	$\frac{20}{49}$		
99.	$\frac{15}{27} - \frac{6}{33}$		
100.	$\frac{6}{20}$		
101.0001	$4\frac{24}{43} - \frac{18}{6}$	.0003	11
102.0002	$4\frac{17}{43} - \frac{6}{49}$	.0007	11
103.0003	$1\frac{6}{43} + \frac{16}{49}$	.0010	4
104.	$\frac{12}{47}$		
105.	$\frac{6}{21}$		
106.0003	$2\frac{36}{41} + \frac{24}{43}$	.0009	9
107.0004	$2\frac{31}{41} - \frac{6}{33}$	.0012	7
108.	$\frac{10}{27}$		
108.9998	$2\frac{16}{39} + \frac{6}{49}$	.0006	7
110.	$\frac{12}{43}$		
111.0001	$3\frac{29}{47} + \frac{17}{49}$	.0003	11
111.9980	$4\frac{10}{31} - \frac{14}{33}$	.0063	11
112.9990	$3\frac{29}{47} - \frac{13}{49}$	.0004	9
113.9996	$1\frac{33}{49} + \frac{24}{43}$	.0014	7
115.	$\frac{6}{23}$		
116.	$\frac{12}{49}$		
117.0001	$7\frac{7}{47} - \frac{6}{10}$	.0002	20
117.9994	$1\frac{6}{49} + \frac{24}{41}$	.0019	5
119.0005	$3\frac{6}{47} - \frac{18}{33}$	.0015	8
120.	$\frac{12}{49}$		
120.9982	$1\frac{14}{47} - \frac{15}{19}$	.0055	3
122.0003	$3\frac{11}{41} - \frac{17}{49}$	.0008	11
123.0006	$1\frac{17}{49} + \frac{17}{49}$	.0018	5
124.	$\frac{10}{41}$		
125.0010	$2\frac{17}{41} - \frac{12}{43}$	.0031	8
125.9985	$3\frac{16}{16} - \frac{7}{20}$	.0047	11
127.0002	$2\frac{24}{49} + \frac{12}{43}$	.0006	9
128.	$\frac{6}{18}$		
128.9997	$5\frac{24}{41} + \frac{16}{18}$	.0011	19
130.	$\frac{12}{47}$		
130.9990	$2\frac{10}{49} + \frac{24}{43}$	.0031	11
132.	$\frac{10}{33}$		

Compound Index Movements on the Nos. 1, 2, 3 and 4, Design 1893, Universal Milling Machines. (Absolute and Approximate.)

NO. OF TEETH.	MOVES.	ERROR ON ONE DIAMETRAL PITCH GEAR.	TIMES AROUND.
133.0006	$3\frac{2}{3}-\frac{14}{33}$	.0020	11
134.0002	$3\frac{7}{27}+\frac{15}{19}$	.0007	13
135.	$\frac{8}{27}$		
136.	$\frac{6}{17}$		
137.0001	$3\frac{17}{33}-\frac{9}{49}$	.0005	11
138.	$\frac{11}{33}-\frac{1}{23}$		
138.9998	$2\frac{15}{37}+\frac{24}{49}$	.0005	11
140.	$\frac{11}{35}$		
141.0007	$1\frac{15}{36}+\frac{22}{49}$	.0022	8
141.9998	$4\frac{1}{17}+\frac{10}{19}$	.0006	15
142.9991	$1\frac{19}{49}-\frac{15}{19}$	.0029	5
144.	$\frac{7}{18}$		
145.	$\frac{8}{29}$		
145.9994	$2\frac{9}{37}-\frac{8}{49}$	.0018	7
147.	$\frac{18}{49}-\frac{3}{49}$		
148.	$\frac{19}{37}$		
149.0003	$3\frac{9}{43}-\frac{8}{49}$	.0010	11
150.	$\frac{7}{15}$		
151.0008	$1\frac{15}{18}-\frac{9}{49}$	.0024	7
152.	$\frac{7}{19}$		
153.0002	$2\frac{17}{19}-\frac{6}{49}$	.0005	11
154.	$\frac{8}{21}-\frac{3}{33}$		
155.	$\frac{7}{15}$		
156.	$\frac{19}{39}$		
157.0003	$2\frac{11}{21}+\frac{8}{33}$	.0011	11
157.9990	$5\frac{8}{23}-\frac{13}{15}$	.0031	19
159.0002	$2\frac{9}{37}+\frac{18}{49}$	.0007	10
160.	$\frac{5}{26}$		
161.0016	$2\frac{19}{19}-\frac{7}{49}$	.0051	9
161.9982	$1\frac{19}{39}-\frac{9}{49}$	.0057	7
162.9996	$3\frac{7}{37}-\frac{24}{49}$	.0013	11
164.	$\frac{19}{41}$		
165.	$\frac{8}{33}$		
165.9989	$1\frac{19}{43}+\frac{18}{49}$	.0035	7
166.9995	$2\frac{8}{23}+\frac{3}{33}$	.0015	9
168.	$\frac{5}{21}$		
169.0005	$1\frac{37}{37}+\frac{11}{18}$	.0016	9
170.	$\frac{4}{17}$		
170.9997	$1\frac{29}{47}+\frac{7}{49}$	.0008	7
172.	$\frac{19}{43}$		
173.0003	$1\frac{7}{43}+\frac{11}{18}$	.0011	6

Compound Index Movements on the Nos. 1, 2, 3 and 4, Design 1893, Universal Milling Machines. (Absolute and Approximate.)

NO. OF TEETH.	MOVES.	ERROR ON ONE DIAMETRAL PITCH GEAR.	TIMES AROUND.
174.	$\frac{11}{33} - \frac{3}{29}$		
174.9964	$1\frac{4}{31} + \frac{4}{33}$	.0112	6
176.0024	$1\frac{13}{33} + \frac{11}{13}$	.0075	7
176.9991	$2\frac{17}{47} + \frac{4}{49}$	.0027	11
177.9995	$3\frac{23}{47} + \frac{11}{43}$	.0014	17
179.0002	$2\frac{23}{47} - \frac{13}{49}$	.0006	11
180.	$\frac{8}{27}$		
180.9906	$2\frac{3}{43} + \frac{12}{43}$	.0012	11
182.	$\frac{8}{39} + \frac{7}{29}$		
183.0003	$1\frac{21}{44} + \frac{8}{49}$	.0000	8
184.	$\frac{8}{23}$		
185.	$\frac{8}{37}$		
186.	$\frac{11}{17} - \frac{11}{33}$		
186.9982	$1\frac{19}{47} + \frac{12}{43}$	.0056	8
188.	$\frac{19}{47}$		
189.0015	$2\frac{29}{47} - \frac{16}{49}$	.0046	11
190.	$\frac{4}{19}$		
191.0015	$1\frac{29}{47} + \frac{11}{49}$	.0046	10
191.9983	$2\frac{21}{47} - \frac{12}{49}$	.0055	11
193.0007	$1\frac{4}{37} - \frac{12}{49}$	.0021	4
193.9981	$2\frac{27}{47} - \frac{16}{49}$	.0061	11
195.	$\frac{8}{39}$		
196.	$\frac{19}{49}$		
196.9996	$1\frac{13}{43} + \frac{14}{49}$	.0013	11
198.	$\frac{3}{27} + \frac{3}{33}$		
199.0005	$2\frac{11}{44} - \frac{4}{49}$	.0014	11
200.	$\frac{4}{20}$		
201.0003	$2\frac{15}{43} + \frac{19}{49}$	.0011	13
201.9991	$3\frac{11}{44} + \frac{8}{49}$	.0028	17
202.9979	$1\frac{13}{43} + \frac{8}{49}$	.0065	9
203.9992	$2\frac{19}{43} + \frac{8}{49}$	.0025	13
205.	$\frac{8}{41}$		
206.0007	$2\frac{13}{43} + \frac{8}{49}$	.0023	15
206.9991	$3\frac{8}{41} - \frac{23}{44}$	.0029	14
207.9980	$1\frac{19}{47} + \frac{16}{49}$	.0063	9
208.9987	$\frac{8}{49} + \frac{8}{41}$	.0041	2
210.	$\frac{4}{21}$		
211.0013	$1\frac{23}{44} + \frac{11}{49}$	.0039	11
211.9995	$3\frac{4}{47} + \frac{8}{49}$	.0017	17
212.9990	$1\frac{13}{43} + \frac{8}{49}$	.0033	8
214.0003	$3\frac{9}{47} - \frac{19}{49}$	.0010	15

Compound Index Movements on the Nos. 1, 2, 3 and 4, Design 1893, Universal Milling Machines. (Absolute and Approximate.)

NO. OF TEETH.	MOVES.	ERROR ON ONE DIAMETRAL PITCH.	TIMES AROUND.
215.	$\frac{8}{13}$		
216.	$\frac{2}{27}$		
217.0014	$2\frac{2}{13} + \frac{16}{19}$	.0044	13
217.9086	$1\frac{22}{27} - \frac{9}{19}$	.0042	7
218.9080	$3\frac{13}{23} - \frac{16}{19}$	.0034	19
220.	$\frac{8}{23}$		
220.9996	$1\frac{4}{27} - \frac{7}{19}$	.0013	6
222.0019	$2\frac{8}{13} - \frac{16}{19}$	.0060	11
222.9998	$2\frac{16}{23} + \frac{14}{19}$	.0005	16
223.9955	$2\frac{9}{23} + \frac{8}{23}$	.0143	13
225.	$\frac{13}{18} - \frac{20}{27}$		
225.9995	$1\frac{33}{36} + \frac{16}{19}$	.0014	13
226.9999	$3\frac{9}{23} + \frac{8}{19}$	.0005	18
228.0010	$2\frac{4}{17} - \frac{14}{19}$	.0032	11
229.0002	$2\frac{18}{27} - \frac{16}{19}$	.0007	12
230.	$\frac{8}{23}$		
231.	$\frac{9}{21} + \frac{8}{33}$		
232.	$\frac{9}{29}$		
233.0007	$1\frac{18}{27} + \frac{8}{19}$	.0022	11
234.0022	$2\frac{21}{23} + \frac{8}{33}$	.0068	17
235.	$\frac{8}{27}$		
236.0007	$2\frac{10}{23} + \frac{8}{19}$	.0021	17
236.9908	$2\frac{18}{27} - \frac{8}{19}$	.0006	13
237.9992	$2\frac{4}{31} + \frac{14}{23}$	.0024	15
238.9997	$1\frac{23}{33} + \frac{14}{19}$	.0008	11
240.	$\frac{3}{18}$		
240.9997	$1\frac{6}{11} + \frac{24}{33}$	.0010	9
241.9996	$2\frac{24}{31} - \frac{8}{27}$	.0013	15
242.9997	$1\frac{22}{27} - \frac{8}{27}$	.0009	10
243.9986	$2\frac{18}{31} + \frac{16}{33}$	.0044	17
245.	$\frac{8}{27}$		
246.0012	$1\frac{8}{13} - \frac{16}{23}$	.0037	5
247.0002	$2\frac{13}{23} - \frac{8}{27}$	.0007	14
248.	$\frac{8}{31}$		
249.0002	$3\frac{4}{13} - \frac{8}{19}$	.0005	19
250.0027	$2\frac{9}{27} - \frac{8}{19}$	.0083	13

If the index head is set for a lower number than the desired one, then the larger of the two moves must carry the top of blank in the opposite direction from that in which the cross-slide is moved, while if the index head is set for a higher number than the desired one, then the larger of the two moves must carry the top of blank in the same direction as that in which the cross-slide is moved."

The foregoing tables were prepared with reference to the No. 1 Universal Milling Machine, as this machine is most frequently used for the operations requiring indexing mechanism. When, however, the movements involve the use of the two outer circles of holes of the index plate, the figures given can be applied to the No. 4 Universal Milling Machine, the crank pin of this machine being used, at such times, in the outer circle of holes, and the back pin in the second circle.

The following table gives a number of movements of the No. 4 machine by the compound method :

Compound Index Movements on the No. 4 Universal Milling Machine.

NO. OF DIVISIONS.	MOVES.	NO. OF DIVISIONS.	MOVES.
57	$\frac{6}{18} + \frac{1}{19}$	217	$\frac{11}{21} - \frac{11}{21}$
93	$\frac{1}{18} + \frac{3}{21}$	228	$\frac{6}{18} - \frac{6}{18}$
114	$\frac{13}{18} - \frac{6}{18}$	282	$\frac{13}{18} - \frac{3}{21}$
141	$\frac{16}{18} - \frac{17}{21}$	285	$\frac{13}{21} - \frac{16}{18}$
171	$\frac{6}{18} - \frac{6}{18}$	304	$\frac{6}{21} - \frac{6}{18}$
186	$\frac{13}{18} - \frac{11}{21}$		

The following table is the result of Mr. Schneider's ingenious suggestion, that sometimes it is not necessary to use the compound indexing for obtaining each division of the work, but that it is often sufficient to use the method once in going around the work twice or twice in going around three times, etc. For example : 96 divisions may be obtained by indexing for 48 and going completely around the work in the ordinary way, then dividing one of the spaces in equal parts, or obtaining one of the desired 96 divisions by the compound method ($\frac{3}{16}$ + $\frac{5}{20}$), and finally by again going around the work, indexing as at first for 48 divisions :

Compound Index Movements for Nos. 1, 2, 3 and 4, Design 1893, Universal Milling Machines (Schneider's Method.)

NUMBER OF TEETH.	SET FOR	TIMES AROUND.	ORDINARY MOVES.	COMPOUND MOVES.	NUMBER OF COMPOUND MOVES.
69	23	3	$1\frac{7}{23}$	$\frac{8}{23} - \frac{11}{44}$	2
87	29	3	$1\frac{11}{29}$	$\frac{12}{29} - \frac{11}{44}$	2
91	13	7	$3\frac{9}{39}$	$\frac{6}{39} + \frac{11}{49}$	6
93	31	3	$1\frac{9}{31}$	$\frac{9}{31} + \frac{11}{44}$	2
96	48	2	$\frac{11}{48}$	$\frac{3}{16} + \frac{5}{20}$	1
138	46	3	$\frac{30}{23}$	$\frac{11}{23} - \frac{5}{23}$	2
174	58	3	$\frac{30}{29}$	$\frac{11}{29} - \frac{5}{20}$	2
182	26	7	$1\frac{21}{39}$	$\frac{9}{39} + \frac{7}{49}$	6
186	62	3	$\frac{30}{31}$	$\frac{11}{31} - \frac{11}{44}$	2
225	45	5	$\frac{24}{27}$	$\frac{9}{27} - \frac{2}{20}$	4
231	21	11	$1\frac{12}{21}$	$\frac{9}{21} + \frac{9}{33}$	10
253	23	11	$1\frac{11}{23}$	$\frac{1}{2} - \frac{11}{46}$	10
259	37	7	$1\frac{27}{37}$	$\frac{11}{37} - \frac{7}{49}$	6
272	136	2	$\frac{7}{17}$	$\frac{10}{17} - \frac{7}{17}$	1
273	39	7	$1\frac{7}{39}$	$\frac{11}{39} - \frac{11}{49}$	6
276	92	3	$\frac{10}{23}$	$\frac{11}{23} - \frac{11}{44}$	2
287	41	7	$\frac{40}{41}$	$\frac{11}{41} - \frac{7}{41}$	6
288	144	2	$\frac{5}{18}$	$\frac{7}{16} - \frac{5}{18}$	1
301	43	7	$\frac{40}{43}$	$\frac{11}{43} - \frac{11}{43}$	6
304	152	2	$\frac{7}{19}$	$\frac{10}{16} - \frac{7}{19}$	1

Adjustable Back Pins.   In his articles upon compounding indexing, Mr. Fred. J. Miller has shown how much the usefulness of an index plate is increased if the back pin of the mechanism is made adjustable, the same as the front pin, and Mr. Miller holds a patent on Index Centres, of which we hold a shop right, which is important in this connection. When such an adjustment is applied to our machines, the number of the attainable movements is increased. The following table is indicative of this:

COMPOUND INDEX MOVEMENTS.

Extra Divisions Obtained by Making the Back Pin Adjustable on Universal Milling Machines.

NO. OF DIVISIONS.	MOVES.	NO. OF DIVISIONS.	MOVES.
51	$\frac{2}{17} + \frac{11}{33}$	204	$\frac{9}{17} - \frac{6}{18}$
57	$\frac{6}{18} + \frac{7}{19}$	207	$\frac{7}{23} - \frac{9}{27}$
63	$\frac{11}{21} + \frac{7}{27}$	217	$\frac{11}{31} - \frac{11}{37}$
102	$\frac{1}{17} + \frac{6}{18}$	222	$\frac{12}{37} - \frac{13}{18}$
111	$\frac{3}{37} + \frac{11}{33}$	228	$\frac{6}{18} - \frac{3}{19}$
114	$\frac{12}{33} - \frac{1}{19}$	246	$\frac{13}{18} - \frac{7}{41}$
123	$\frac{36}{18} - \frac{11}{41}$	252	$\frac{7}{41} + \frac{9}{27}$
126	$\frac{2}{41} + \frac{6}{27}$	255	$\frac{6}{18} - \frac{7}{17}$
129	$\frac{13}{18} - \frac{7}{43}$	258	$\frac{11}{21} - \frac{13}{18}$
141	$\frac{36}{18} - \frac{15}{47}$	261	$\frac{7}{17} - \frac{9}{29}$
153	$\frac{12}{17} - \frac{1}{18}$	279	$\frac{9}{27} + \frac{7}{41}$
161	$\frac{5}{23} - \frac{9}{21}$	282	$\frac{13}{18} - \frac{7}{47}$
171	$\frac{6}{18} - \frac{1}{19}$	285	$\frac{6}{19} - \frac{6}{18}$
189	$\frac{6}{21} - \frac{2}{27}$	306	$\frac{6}{17} - \frac{1}{18}$
192	$\frac{9}{18} - \frac{3}{19}$		

Mr. Miller has given a great deal of attention to the results obtained by the use of two adjustable pins in connection with an index plate, and his tables are interesting and, in many instances, would be of great value. We do not reproduce them, however, as none happen to be applicable to the index plates regularly furnished with the machines.

BEVEL GEARS.

CUTTING BEVEL GEARS IN A UNIVERSAL MILLING MACHINE.

An article written by Mr. O. J. Beale appeared in the *American Machinist*, June 20, 1895, with the above title and covers this subject most thoroughly. The following is a reprint of this article, together with the illustrations :

" Bevel gears connect shafts whose axes meet when sufficiently prolonged. The teeth of bevel gears are formed about the frustrums of cones whose apexes are at the same point where the shafts meet. In Fig. 63 we have the axes A O and B O, meeting at O, and the apexes of the two cones are also at O. If, in any bevel gear, the teeth were sufficiently prolonged toward the apex, they would become infinitely small ; that is, the teeth would all end in a point, or vanish at O. We can also consider a bevel gear as beginning at the apex and becoming larger and larger as we go away from the apex.

" Fig. 64 is a section of a pair of bevel gears, the gear C D being twice as large as C I. The outer surface of a tooth, J E, Fig. 64, is called its face. The distance C c is called the length of the face of tooth, which is often designated by the letter F : strictly, the distance J E is longer than C c, but the difference is not usually recognized. The outer part of a tooth at C is called its large end and the inner part C the small end. In speaking of the pitch of bevel gears we always name that at the large end or at the large pitch circle, thus the pitch of a bevel gear, at its large pitch circle, corresponds to that of a spur gear. The sizes of the teeth at the small end are in the same proportion to those at the large end as the distance O C is to the apex distance O C, that is, we can figure the sizes as we would figure the thickness of a wedge at any point, after we know the length and the thickness at the

butt end.   There are convenient tables for spur gear teeth
in Brown & Sharpe's ' Practical Treatise on Gearing.'

"Nothing is gained by having the length of face J E
longer than five times the tooth thickness at the large pitch
circle, and even this is too long when it is more than a
third of the apex distance C O.   To cut a bevel gear with
a rotary cutter, as in Fig. 65, is at best but a compromise,
because the teeth change pitch from end to end, so that
the cutter, being of the right shape for the large ends of
the teeth, cannot be right for the small ends, and the variation is too great when the length of face is longer than a
third of the apex distance. Frequently the teeth have to be
be rounded over at the small ends by filing; the longer
the face the more we have to file.

" The data for bevel gears can be figured by trigonome-
try, and without a drawing, as in Brown & Sharpe's ' Formulas in Gearing; ' they can also be obtained by the help
of measurement of a drawing. We will suppose that we
have a drawing, and that we have the gear blanks turned
correctly enough to gauge from in making our settings for
cutting the teeth.   It is possible to cut a good gear from a
blank somewhat incorrectly turned, but it is quite incon-
venient to make the settings without guiding by the
blank.

" These data are needed before beginning to cut :

" The pitch and the number of teeth, the same as for
spur gears.

"The number of the cutter, so as to select one of correct
form,   Brown & Sharpe have a system of eight cutters for
each pitch. A pair of bevel gears having different numbers of teeth may require two cutters.

"The whole depth of the tooth spaces both at the outer
and inner ends, which can be designated by $D'' + f$ at the
outer end, and by $D''' + f$ at the inner end.

" The thickness of the teeth at the outer and inner pitch
lines, which we will designate by t at the outer, and by t'
at the inner end.

"The heights of the teeth above the two pitch lines, s at the outer end and s' at the inner.

"The cutting angles, or the angles that the path of the cutter makes with the axes of the gears.   In Fig. 64 the cutting angle for the gear C D is A O G, and the cutting angle for the pinion is B O H.   In Fig. 66 a cutter is shown passing through a tooth space along the line O G, the gear being set to the cutting angle A O G.

"The lines G O and H O. Fig. 64, are called working depth lines, because they show the depth that the teeth of the two gears engage.   The spaces are cut deeper than these lines an amount which is called the clearance or f; this f or clearance is the same at both ends of the teeth, when cut with a rotary cutter.

"Brown & Sharpe have kindly offered to turn up a pair of blanks and cut them, while I note down the machine settings.   I choose gears of 8 pitch, 24 and 12 teeth, $\frac{1}{2}$ inch face, shown in Figs. 64 and 72.

"The shape of the teeth of one of these gears differs so much from that in the other gear that two cutters are required.   The cutters may be determined as follows: Twice the length of the line C K (Fig. 64) in inches, per-pendicular to C O, multiplied by the diametrial pitch, equals the number of teeth for which to select a cutter, as to form, to cut the twelve-tooth gear.   This number is about 13, indicating a No. 8 cutter.   In the same way, multiplying twice the corresponding line in the other gear by the diametrial pitch, we have about 54, calling for a No. 3 cutter.   C K is sometimes called the back cone radius.

"This way of selecting cutters is based upon the idea of shaping the teeth as near right at the large end as practi-cable, and then to file the small ends where the cutter has not rounded them over enough.   In Fig. 69 the tooth L has been cut to thickness at both pitch lines, but it must still be rounded at the inner end, or at the part corres-ponding to J in Fig. 64.   The teeth M M, Fig. 69, have been filed.   In thus rounding the teeth we must avoid making them any thinner at the pitch line.

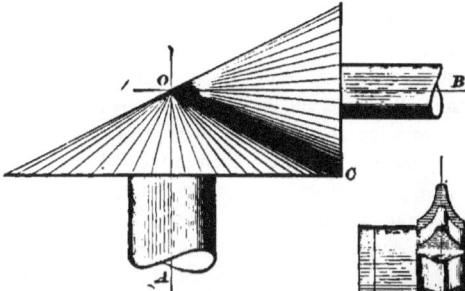
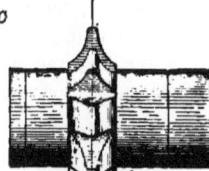

Fig. 63.

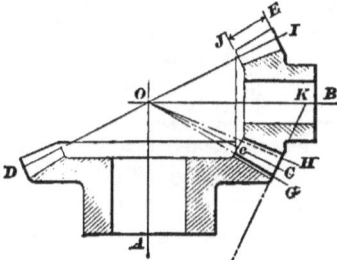

Fig. 64.

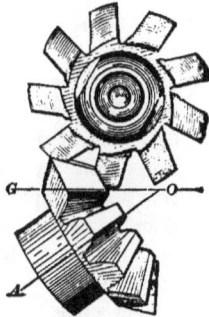

Fig. 65.

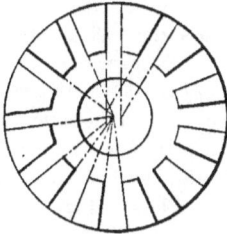
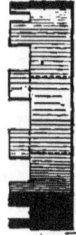

Fig. 67.

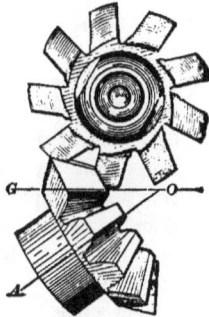

Fig. 66.

" In cutting a bevel gear the finished spaces are not
always of the same form as the cutter might be expected
to make, because of the changes in the position of the
gear blank in order to cut the two sides of the spaces.
The cutter, of course, being thin enough for the small end
of a space, the large end of a space is cut to the required
width by rotating the blank and adjusting sidewise, and
we usually cut twice through each space.   Thus in Fig. 65
a gear is in position to have a space widened at the large
end e, and the last chip to be taken off the tooth on the
right of the cutter, the blank having been moved to the
right and then rotated in the direction of the arrow.   It
may be well to remember that in setting to finish the side
of a tooth the gear blank is moved sidewise in the direction to take this tooth away from the cutter, and then the
blank is rotated by indexing the spindle to bring the tooth
up against the cutter.   Now this tends not only to cut the
space wider at the largest pitch circle, but also to cut
still more off at the face of the tooth ; that is, the teeth
may be cut rather thin at the face, and left rather thick at
the roots.   This tendency is greater as a cutting angle B
O H, Fig. 64, is smaller or as a bevel gear is more like a
spur gear, because when the cutting angle is small the
blank must be rotated through a greater arc in order to
set to cut the right tooth thickness at the outer pitch circle.   This can be understood by Figs. 67 and 70; in Fig.
67 the teeth are cut square across the axis, and the rotation of the blank in order to cut spaces wider at the outside has not narrowed the faces of the teeth any more
than the roots.   The different positions of the cutter in both
the unfinished and finished spaces is shown by the dotted
lines.   Now as the cutting angle of a bevel gear approaches
a right angle, as in Fig. 67, we find less and less tendency
to cut the tooth faces too narrow.   In Fig. 70 a cutter is
passing through a space in a spur gear, which has no
cutting angle ; it is clear that any rotation of this blank
tends directly to change the shape of the teeth as regards
the thickness of the face and the root.   In this particular

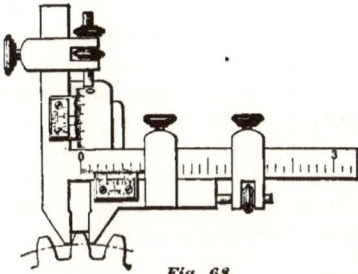

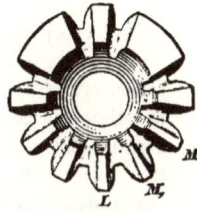

Fig. 69.

Fig. 68.

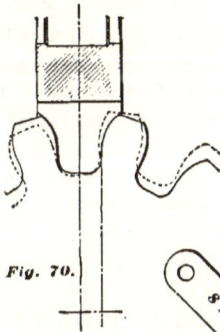

Fig. 71.

Fig. 70.

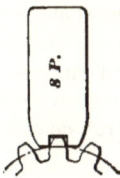

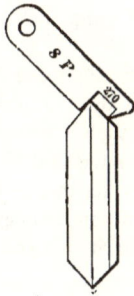

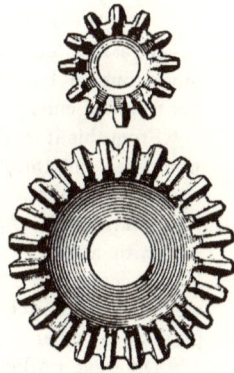

Fig. 75.   *Fig. 74.*   *Fig. 72.*

setting, shown by the centre lines of the cutter and of the gear, the roots are not changed while the faces are made considerably thinner, as indicated by the dotted lines. This change in the shape of the spaces caused by the rotation of the blank may be so great as to require the substitution of a cutter that is narrower at e e', Fig. 65, that is, a cutter for cutting a higher-numbered gear. In using the cutter for a higher-numbered gear the radius of curvature of the tooth sides is lengthened, but the longer tooth curves are not so objectionable as the thin tooth faces.

"In the order for cutting the gears, Fig. 72, the foreman of the gear department calls for a No. 6 cutter for 17 to 20 teeth to cut the 12-tooth gear, instead of a No. 8, as I have just figured. This is a little surprising, but it is still more surprising to be told by the foreman of the cutter department that he would fill an order for a cutter to cut this same gear by sending a No. 8 for 12 to 13 teeth. I told him about the gear order, and he replied, 'I don't know what they do over there, but I should think they work by guess. I make cutters according to the rule in the catalogue. If there is a better rule I should like to have it. We seldom or never have any fault found with bevel gear cutters because they round the teeth over too much, but we have had complaints because the teeth are not rounded enough.'

"From this it might be supposed that Brown & Sharpe would use one thing on their own work, and send their customer another, but such a supposition would not be fully warranted. As I have said, the cutting of a bevel gear with a rotary cutter, as in Fig. 65, is a compromise. The necessity for a compromise is evident when we consider that an 8-pitch gear at one end may run down to 12 pitch at the other.

" Different workmen prefer to compromise in different ways. A workman can avoid much filing of the teeth by the use of the No. 8 cutter, but the tooth faces will be considerably too thin at the large ends. He can also avoid

some filing by an extra cut upon each side of the teeth at the small ends, making four cuts in all. I have known this compromise with a coarse pitch and long face. If the faces are short and the pitch is fine he can sink the cutter below the regular depth at the outside, and cut only once around; this compromise is little liked, and is seldom employed ; it may require a special form and thickness of cutter. The best compromise is to cut twice around, shaping the teeth as nearly correct as practicable at the large ends, and then file the small ends in low-numbered gears, or those of fewer than twenty teeth. A workman can soon acquire the skill to file the teeth almost as perfect as they can be planed from a template. Still most workmen prefer not to file the 12-tooth gear, Fig. 64, which is the reason why the rule in the catalogue is followed in filling an order for a gear cutter.

" In the selection of a cutter, the foreman of the cutting department tells me that he cannot give a definite rule, and that he is open to the criticism of guessing. If a pinion of only 12 teeth is to run with a gear two or more times as large, he takes a cutter shaped for 17 to 20 teeth, instead of the usual 12 to 13 teeth. For any gears higher than 25 teeth he would adhere to the rule, and take the cutter indicated by the length of the line C K, Fig. 64. If he has many gears of the same size, he cuts one pair and files them to run together, in order to decide definitely as to the cutters.

" The sizes of the tooth parts at the large end are copied right from a table of spur gear teeth. The distance O c, Fig. 64, is seven-tenths the apex distance O C, so that the sizes of the tooth parts at the small end, all except f, or the clearance, are seven-tenths the large. The order goes to the workman in this form, P standing for diametral pitch, and N for the number of teeth :

FIG. 73.

LARGE GEAR.

$$P = 8$$
$$N = 24$$
$$D'' + f = .270'' \quad D''' + f = .195''$$
$$t = .196'' \quad t' = .137''$$
$$s = .125'' \quad s' = .087''$$
$$\text{Cutting angle} = 59° \; 10'.$$

SMALL GEAR.

$$N = 12.$$

Cutting angle $= 22° \; 18'$.

Cutters, Nos. 3 and 6, 8 P, bevel.

" Before beginning to cut, the bed S, Fig. 73, is set to zero. The dial pointer of the cross feed screw T is set and noted, so that we can adjust the blank to any required distance out of centre with the cutter. The spiral head N is moved up to the cutting angle, which for our 24-tooth gear is 59° 10'; we guess at the $\frac{1}{6}$ degree more than 59 degrees.

" Mark the depth of cut at the outside as in Fig. 74. It is also well enough to mark the depth at the inside, as a check. The thickness of the teeth at the outside is conveniently determined by the solid gauge, Fig. 75. The vernier caliper, Fig. 68, will measure different sizes. If we do not have the vernier caliper, a gauge like Fig. 75, filed to the thickness at the small ends, will answer.

" The index having been set to divide to the right number, our workman cuts two spaces central with the blank, leaving a tooth between that is a little too thick, as in the upper part of Fig. 69. The tooth has to be cut away more in proportion from the large than from the small end, which is the reason for setting the blank out of centre, as in Fig. 65.

" It is important to remember that the part of the cutter that is finishing one side of a tooth at the pitch line should be central with the gear blank, in order to know at once in which direction to set the blank out of centre.

We cannot readily tell how much out of centre to set the
blank until we have cut and tried, because the same part
of a cutter does not cut to the pitch line at both ends of
the tooth.   As a trial distance out of centre, we can take
about one-tenth to one-eighth of the thickness of the
teeth at the large end.   The actual distance out of centre
for the 12-tooth gear is .021 inch, and for the 24-tooth
.030 inch, when using the cutters listed in the catalogue.

"After a little practice a workman can set his blank the
trial distance out of centre, and take his first cuts, without
making any central cuts at all, but it is safer to take central cuts like the upper ones in Fig. 69.   The depth of the
cuts is controlled by the shaft U, Fig. 73.   Now, by
means of the screw T, set out of centre the trial distance,
which can be one-tenth the thickness of the tooth at the
large end in a 12-tooth gear, and from that to one-eighth
the thickness in a 24-tooth gear and larger.   The direction out of centre is indicated in Fig. 65, which is to move
the tooth away from the cutter, by turning the cross-feed
screw T, Fig. 73 ; the tooth is then rotated up against the
cutter in the direction of the arrow, Fig. 65, by turning
the index crank R, Fig. 73, just enough to trim the side
of the tooth as shown in Fig. 65.   The blank is now set
the same distance out of centre in the other direction,
rotated contrary to the arrow, Fig. 65, and the other side
of the tooth is trimmed until one end is nearly down to the
right thickness.   If now the thickness of the small end is
in the same proportion to the large end as O c is to O C
in Fig. 64, we can at once trim the tooth to the right
thickness by turning the index crank R, Fig. 73.   The
object of setting out of centre is to trim more from the
large end e', Fig. 65 ; if now we find that by our two settings the tooth is still going to be too thick at e', when the
small end is right, the out of centre distance must be
increased.   .

"An easy way to remember this principle is this : Too
much out of centre leaves the small end of the tooth too

thick, while too little out of centre leaves the small end too thin.

".After the proper distance out of centre has been learned, the teeth can be finish-cut by going around out of centre, first on one side, and then on the other, without cutting any central spaces at all. If, however, a gear is coarser than 5-pitch diametral, it is sometimes well enough to cut all the spaces central at first.

" Blanks are not always turned nearly enough alike to be cut without a different setting of the machine for different blanks. If the hubs vary in length, the height of the knee Q, Fig. 73, has to be varied. In thus varying, the same depth of cut or the exact $D'' + f$ may not always be reached. A slight difference in the depth is not so objectionable as the incorrect tooth thickness that it may cause, hence it is well, after cutting once around and finishing one side of the teeth, to adjust the spindle N, by means of the index crank R, until the right tooth thickness is obtained, paying more attention to the tooth thickness than to the number of index spaces covered by the crank. Of course, if the blanks are alike, it is easier to count the index spaces in passing from one side of the tooth to the other.

"After a gear is cut and before it is filed it is sometimes not a very satisfactory looking piece of work. In Fig. 69 the tooth L is as the cutter left it, and is ready to be filed to the shape of the teeth M M."

USE OF MILLING MACHINES.

CUTTING SPIRALS WITH UNIVERSAL MILLING MACHINES.

Selection of Change Gears. The indexing head stock or spiral head, as indicated in connection with the descriptions of the Univeral Milling Machines, is used for cutting spirals, the flutes of twist drills, for example, as well as for indexing or dividing. A positive rotary movement is given to the work while the spiral bed is being moved lengthways by the feed screw, and the velocity ratios of these movements are regulated by four change gears, shown in position in Fig. 5, and known as the gear on worm or worm gear, first gear on stud, the first gear put on stud, second gear on stud and gear on screw or screw gear. The screw gear and first gear on stud are the drivers and the others the driven gears. Usually these gears are of such ratio that the work is advanced more than an inch while making one turn and thus the spirals, cut on Milling Machines, are designated in terms of inches to one turn, rather than turns, or threads per inch; for instance a spiral is said to be of 8 inches lead, not that its pitch is ⅛ turn per inch.

The feed screw of the spiral bed has four threads to the inch, and forty turns of the worm make one turn of the spiral head spindle; accordingly, if change gears of equal diameter are used, the work will make a complete turn while it is moved lengthways 10 inches; that is the spiral will have a lead of 10 inches. But this lead is practically the lead of the machine, as it is the resultant of the action of the parts of the machine that are always employed in this work, and is so regarded in making the calculations used in cutting spirals.

In principle, these calculations are the same as for change gears of a Screw Cutting Lathe. The compound ratio of the driven to the driving gears equals in all cases, the ratio of the lead of the required spiral to the lead of the machine. And this can be readily understood by changing the diameters of the gears.

Gears of the same diameter produce, as explained above, a spiral with a lead of 10 inches, which is the same lead as the lead of the machine. Three gears of equal diameter and a driven gear double this diameter produce a spiral with a lead of 20 inches, or twice the lead of the machine · and with both driven gears twice the diameters of the drivers, the ratio being compound, a spiral is produced with a lead of 4o inches or four times the machine's lead. Conversely, driving gears twice the diameter of the driven, produce a spiral with a lead equal to $\frac{1}{4}$ the lead of the machine or $2\frac{1}{2}$ inches.

Expressing the ratios as fractions, $\frac{\text{Driven Gears}}{\text{Driving Gears}} = \frac{\text{Lead of Required Spiral}}{\text{Lead of Machine}}$ or as the product of each class of gears determines the ratio, the head being double geared, and as the lead of the machine is ten inches $\frac{\text{Product of Driven Gears}}{\text{Product of Driving Gears}} = \frac{\text{Lead of Required Spiral}}{10}$ That is, the compound ratio of the Driven to the Driving Gears may always be represented by a fraction whose numerator is the lead to be cut and whose denominator is ten. Or, in other words, the ratio is as the required lead is to 10, that is, if the required lead is 20 the ratio is 20 : 10, or to express this in units instead of tens, the ratio is always the same as one tenth of the required lead is to one. And frequently this is a very convenient way to think of the ratio; for example, if the ratio of the lead is 40, the gears are 4 : 1. If the lead is 25, the gears are 2.5 : 1, etc.

To illustrate the usual calculations, assume as in Fig. 5 that a spiral of 12 inch lead is to be cut. The compound ratio of the driven to the driving gears equals the desired lead divided by 10, or it may be represented by the fraction $\frac{12}{10}$. Resolving this into two factors to represent the two pairs of change gears, $\frac{12}{10} = \frac{3}{2} \times \frac{4}{5}$. Both terms of the first factor are multiplied by such a number (24 in this instance) that the resulting numerator and denominator will correspond with the number of teeth of two of the change gears furnished with the machine, (such multiplications not affecting the value of a fraction) $\frac{3}{2} \times \frac{24}{24} = \frac{72}{48}$. The second factor is similarly treated $\frac{4}{5} \times \frac{8}{8} = \frac{32}{40}$, and the gears with 72 and 32 and 48 and 40 teeth are selected, $\frac{12}{10} = \left(\frac{72 \times 32}{48 \times 40}\right)$. The first two are the driven, and the last two the drivers, the numerators of the fractions having represented the driven gears, and the 72 is placed as the worm gear, the 40 as the first on stud, 32 the second on stud and 48 as the screw gear. The two driving gears might be transposed and the two driven gears might also be transposed without changing the spiral. That is, the 72 could be used as the second on stud and the 32 as the worm gear, if such an arrangement was more convenient.

From what has been said, the rules are plain :

Rules for obtaining Ratio of the Gears necessary to Cut a Given Spiral. Note the ratio of the required lead to ten. This ratio is the compound ratio of the driven to the driving gears. Example: if the lead of required spiral is 12 inches, 12 to 10 will be the ratio of the gears.

Or, divide the required lead by 10 and note the ratio between the quotient and 1. This ratio is usually the most simple form of the compound ratio of the driven to the driving gears. Example: if the required lead is 40 inches, the quotient $40 \div 10$ is 4 and the ratio 4 to 1.

Rule for Determining Number of Teeth of Gears required to Cut a Given Spiral. Having obtained the ratio between the required lead and ten by one of the preceding rules, express the ratio in the form of a fraction; resolve this fraction into two factors, raise these factors to higher terms that correspond with the teeth of gears that can be conveniently used. The numerators will represent the driven and the denominators, the driving gears that produce the required spiral. For example, what gears shall be used to cut a lead of 27 inches?

$$\frac{27}{10} = \frac{3}{2} \times \frac{9}{5} = \left(\frac{3}{2} \times \frac{16}{16}\right) \times \left(\frac{9 \times 8}{5 \times 8}\right) = \frac{48 \times 72}{32 \times 40}$$

From the fact that the product of the driven gears divided by the product of the drivers equals the lead divided by ten, or one-tenth of the lead, it is evident that ten times the product of the driven gears divided by the product of the drivers will equal the lead of the spiral. Hence the rule :

Divide ten times the product of the driven gears by the product of the drivers and the quotient is the lead of the resulting spiral in inches to one turn. For example, what spiral will be cut by gears, with 48, 72, 32 and 40 teeth, the first two being used as driven gears? Spiral to be cut equals $\frac{10 \times 48 \times 72}{32 \times 40} = 27$ inches to one turn.

Rule for Ascertaining what Spiral may be cut by any given Change Gears.

This rule is often of service in determining what spirals may be cut with the gears the workman chances to have at hand.

The tables on pages 218 to 221 give the leads of spirals produced by the gears furnished with the machines.

The chapter on " Continued Fractions " in our Practical Treatise on Gearing is of assistance in selecting gears for fractional spirals.

The change gears having been selected, the next step in cutting spirals is to determine the position at which the spiral bed must be placed to bring the spiral in line with the cutter as the work is being milled, for, if the spiral is not in line with the cutter the groove will not be cut to the shape of the cutter.

Position of the Spiral Bed in Cutting Spirals.

The correct position of the spiral bed is indicated by the angle shown at A, Fig. 77, and this angle, as may be noted from that figure, has the same number of degrees as the angle B, which is termed the angle of the spiral, and is formed by the intersection of the spiral and a line parallel with the axis of the piece being milled. The reason the angles A and B are alike is that their corresponding sides are perpendicular to each other.

The angle of the spiral depends upon the lead of the spiral and the diameter of the piece to be milled. The greater the lead of a spiral of any given diameter, the smaller the angle, and the greater the diameter of any spiral the greater the spiral angle.

The angle may be ascertained in two ways, graphically, or more conveniently by a simple calculation and reference to the accompanying tables. In determining it graphically, a right angle triangle is drawn to scale. One of the sides which forms the right angle represents the lead of the spiral in inches; the other side represents the circumference of the piece in inches, and the hypothenuse represents the line of the spiral. The angle between the lines representing the line of the spiral and the lead of the spiral is the angle of the spiral. This angle can be transferred from the drawing to the work by a bevel protractor, or even by cutting a paper templet and winding it about the work as shown in Fig. 78. The machine is then set so that the spiral or groove as it touches the cutter will be in line with the cutter. Or the angle may be measured and the spiral bed set to a corresponding number of degrees by the graduations on the clamp bed.

The natural tangent of the angle of the spiral is the quotient of the circumference of the piece divided by the lead of the spiral. Accordingly, the second method of obtaining the angle of the spiral is to divide the circumference of the piece by the lead, and note the number of degrees opposite the figures that correspond with the quotient in the accompanying tables of natural tangents, pages 222 and 223. The angle having been thus obtained, the spiral bed is set by the graduations on the clamp bed.

Tables, pages 218 to 221, give the lead of spirals produced by the various combinations of the change gears furnished with the machine and the angles of these spirals for diameters from ⅛ inch to 4 inches.

Use of Machines in Cutting Spirals. Before a spiral is cut, it is well to let the mill just touch the work, then run the work along by hand and make a slight spiral mark, and by this mark see whether the change gears give the right lead. The saddle can then be set to the proper angle.

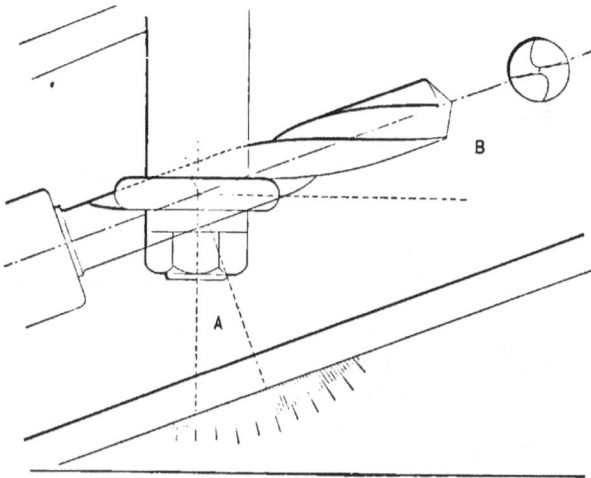

FIG. 77.

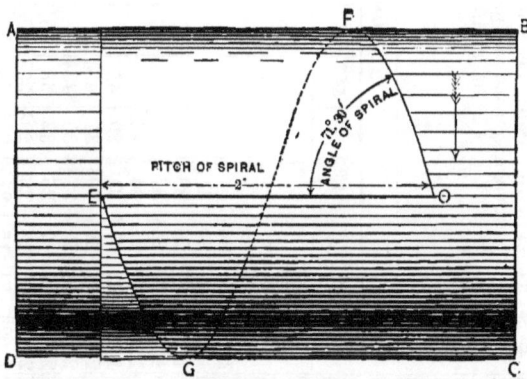

FIG. 78.

TABLE OF CHANGE GEARS, SPIRALS AND ANGLES.

To Find the Angle of Spiral, Divide the Circumference by the Pitch or Lead, and the Quotient will be the Tangent of the Angle. Then Find the Angle in a Table of Tangents. For Complete Explanation See Practical Treatise on Gearing.

Gear on Worm.	First Gear on Stud.	Second Gear on Stud.	Gear on Screw.	Pitch in Inches to one turn.	Diameter of the Blank, or Spiral, to be Cut.																			
					⅛″	¼″	⅜″	½″	⅝″	¾″	⅞″	1″	1¼″	1½″	1¾″	2″	2¼″	2½″	2¾″	3″	3¼″	3½″	3¾″	4″
24	86	24	100	.67	30¼°																			
24	86	28	100	.78	26	44½°																		
24	86	32	100	.89	23½	41																		
24	86	40	100	1.12	19	34½																		
24	86	48	100	1.34	16	30¼	41½°																	
24	64	28	72	1.46	14¾	28	38½																	
24	86	56	100	1.56	13¾	26½	37																	
24	64	32	72	1.67	12¾	25	34¾	43¼°																
32	64	28	72	1.94	11¼	21¾	31	39	45°															
24	64	40	72	2.08	10¼	20½	29½	37	43¼															
32	56	28	72	2.22	9¾	19¼	27½	35	41¼															
24	64	48	72	2.50	8¾	17	25	32	38	43¼°														
40	56	28	72	2.78	8	15½	23	29½	35¼	40½	44¾°													
24	64	56	72	2.92	7½	15	21¾	28¼	34	39	43¼													

Angle for Setting the Saddle.

TABLE OF CHANGE GEARS, SPIRALS AND ANGLES.

Gear on Worm.	First Gear on Stud.	Second Gear on Stud.	Gear on Screw.	Pitch in inches to one turn.	Diameter of the Blank, or Spiral, to be Cut.																		
					1/8″	1/4″	3/8″	1/2″	5/8″	3/4″	7/8″	1″	1¼″	1½″	1¾″	2″	2¼″	2½″	2¾″	3″	3¼″	3½″	3¾″	4″
40	48	28	72	3.24	6¾°	13¼°	19¾°	25¾°	31¼°	36°	40½°	44¼°											
40	48	32	72	3.70	6	11¾	17½	23	28	32½	36½	40½											
56	48	24	72	3.89	5½	11¼	16¾	22	26¾	31¼	35¼	39											
40	72	48	64	4.17	5¼	10½	15¾	20½	25¼	29½	33½	37	43¼°										
48	40	32	86	4.46	4¾	9¾	14¾	19¼	23¾	27¾	31½	35	41½										
40	64	56	72	4.86	4½	9	13½	17¾	22	25¾	29½	33	39	44¼°									
48	40	32	72	5.33	4	8¼	12¼	16½	20¼	23¾	27¼	30½	36½	41½									
56	40	28	72	5.44	4	8	12	16	20	23½	26¾	30	36	41									
56	40	28	64	6.12	3½	7¼	11	14½	17¾	21	24¼	27	33	37¾	42°								
56	40	32	72	6.22	3½	7	10¾	14¼	17½	20¾	23¾	26¾	32½	37¼	41½								
56	48	40	72	6.48	3¼	6¾	10¼	13½	16¾	20	23	25¾	31½	36¼	40¼	44¼°							
64	48	28	56	6.67	3¼	6½	10	13¼	16½	19½	22½	25¼	30¾	35¼	39½	43½							
56	48	40	64	7.29	3	6¼	9¼	12¼	15	18	20½	23½	28½	33	37	41	44¼°						
64	48	40	72	7.41	3	6	9	12	14¾	17¾	20¼	22¾	28¼	32½	36½	40¼	43¾						
64	48	32	56	7.62	2¾	5¾	8¾	11½	14½	17¼	19¾	22¼	27½	32	36	39½	43						
48	32	40	72	8.33	2½	5¼	8	10½	13¼	15¾	18¼	20½	25½	29½	33½	37	40½	43½°					
86	48	28	56	8.95	2½	5	7½	10	12½	14¾	17	19¼	24	28	31¾	35¼	38½	41¼	44°				
56	40	48	72	9.33	2¼	4¾	7¼	9½	11¾	14	16¼	18½	23	27	30½	34	37¼	40¼	43				

Angle for Setting the Saddle.

TABLE OF CHANGE GEARS, SPIRALS AND ANGLES.

DIAMETER OF THE BLANK, OR SPIRAL, TO BE CUT.

ANGLE FOR SETTING THE SADDLE.

Pitch in inches to one turn.	Gear on Screw.	Second Gear on Stud.	First Gear on Stud.	Gear on Worm.
9.52	56	40	48	64
10.29	56	32	40	72
10.37	72	56	48	64
10.50	64	56	40	48
10.67	72	40	40	64
10.94	64	48	32	56
11.11	72	40	72	64
11.66	72	64	32	56
12.00	48	32	48	72
13.12	64	48	40	56
13.33	72	32	32	72
13.71	72	48	48	56
15.24	56	48	28	64
15.56	72	48	48	64
15.75	72	56	32	64
16.87	40	72	64	56
17.14	64	48	32	72
18.75	48	40	48	72

TABLE OF CHANGE GEARS, SPIRALS AND ANGLES.

Gear on Worm.	First Gear on Stud.	Second Gear on Stud.	Gear on Screw.	Pitch in inches to one turn.	Diameter of the Blank, or Spiral, to be Cut.																			
					⅛″	¼″	⅜″	½″	⅝″	¾″	⅞″	1″	1¼″	1½″	1¾″	2″	2¼″	2½″	2¾″	3″	3¼″	3½″	3¾″	4″
72	32	48	56	19.29	1°	2¼°	3½°	4½°	5¾°	7°	8°	9¼°	11½°	13¾°	16°	18¼°	20¼°	22¼°	24°	26°	28°	29¾°	31½°	33°
64	28	48	56	19.59	1	2¼	3¼	4½	5¾	6¾	8	9¼	11½	13½	15¾	18	20	22	23¾	25¾	27½	29¼	31	32¾
72	32	56	64	19.69	1	2¼	3¼	4½	5¾	6¾	8	9	11½	13½	15¾	17¾	20	21¾	23¾	25½	27½	29¼	31	32½
72	24	40	56	21.43	1	2	3¼	4¼	5½	6¼	7½	8½	10½	12½	14½	16½	18½	20¼	22	23¾	25½	27¼	29	30¼
72	28	56	64	22.50	1	2	3	4	5	6	7	8	10	12	13¾	15¾	17½	19¼	21	22¾	24½	26	27¾	29¼
64	32	56	48	23.33	1	2	3	4	5	5¾	6¾	7¾	9¾	11½	13¼	15¼	17	18¾	20¼	22	23½	25¼	27	28¼
72	24	56	64	26.25	1	1¾	2¾	3½	4¼	5	6	7	8½	10¼	12	13½	15	16¾	18¼	19¾	21¼	22¾	24¼	25½
64	28	56	48	26.67	¾	1¾	2¾	3½	4¼	5	6	6¾	8½	10	11¾	13¼	14¾	16½	18	19½	21	22¼	23¼	25¼
64	32	56	40	28.00	¾	1¾	2½	3¼	4	4¾	5¾	6½	8	9½	11¼	12¾	14¼	15¾	17¼	18¾	20	21½	22¾	24
72	28	48	40	30.86	¾	1½	2¼	3	3¾	4½	5	5¾	7¼	8¾	10	11½	13	14¼	15½	17	18½	19½	21	22
72	32	56	40	31.50	¾	1½	2¼	3	3½	4¼	5	5¾	7¼	8½	10	11¼	12¾	14	15¼	16½	18	19¼	20½	21¾
72	32	64	40	36.00	½	1¼	2	2½	3¼	3¾	4½	5	6¼	7½	8¾	10	11	12¼	13½	14¾	16	17	18¼	19¼
72	28	64	40	41.14	½	1	1¾	2¼	2¾	3¼	4	4½	5½	6½	7¾	8¾	9¾	10¾	11¾	13	14	15	16	17
72	28	56	32	45.00	½	1	1½	2	2½	3	3½	4	5	6	7	8	9	10	11	11¾	12¾	13¾	14¾	15½
72	24	64	40	48.00	½	1	1½	2	2½	3	3¼	3¾	4¾	5½	6½	7½	8½	9¼	10¼	11¼	12	13	13¾	14½
72	28	64	32	51.43	½	¾	1½	1¾	2¼	2¾	3	3½	4½	5¼	6	7	7¾	8¾	9½	10½	11¼	12	12¾	13¾
72	24	64	32	60.00	½	¾	1¼	1½	2	2¼	2½	3	3¾	4½	5¼	6	6¾	7½	8¼	9	9½	10¼	11	11¾
72	24	64	28	68.57	¼	¾	1	1¼	1¾	2	2¼	2½	3¼	4	4¼	5¼	5¾	6½	7¼	8	8½	9	9¾	10¼

Angle for Setting the Saddle.

NATURAL TANGENT.

Deg.	0′	10′	20′	30′	40′	50′	60′	
0	.00000	.00290	.00581	.00872	.01163	.01454	.01745	89
1	.01745	.02036	.02327	.02618	.02909	.03200	.03492	88
2	.03492	.03783	.04074	.04366	.04657	.04949	.05240	87
3	.05240	.05532	.05824	.06116	.06408	.06700	.06992	86
4	.06992	.07285	.07577	.07870	.08162	.08455	.08748	85
5	.08748	.09042	.09335	.09628	.09922	.10216	.10510	84
6	.10510	.10804	.11099	.11393	.11688	.11983	.12278	83
7	.12278	.12573	.12869	.13165	.13461	.13757	.14054	82
8	.14054	.14350	.14647	.14945	.15242	.15540	.15838	81
9	.15838	.16136	.16435	.16734	.17033	.17332	.17632	80
10	.17632	.17932	.18233	.18533	.18884	.19136	.19438	79
11	.19438	.19740	.20042	.20345	.20648	.20951	.21255	78
12	.21255	.21559	.21864	.22169	.22474	.22780	.23086	77
13	.23086	.23393	.23700	.24007	.24315	.24624	.24932	76
14	.24932	.25242	.25551	.25861	.26172	.26483	.26794	75
15	.26794	.27106	.27419	.27732	.28046	.28360	.28674	74
16	.28674	.28989	.29305	.29621	.29938	.30255	.30573	73
17	.30573	.30891	.31210	.31529	.31850	.32170	.32492	72
18	.32492	.32813	.33136	.33459	.33783	.34107	.34432	71
19	.34432	.34758	.35084	.35411	.35739	.36067	.36397	70
20	.36397	.36726	.37057	.37388	.37720	.38053	.38386	69
21	.38386	.38720	.39055	.39391	.39727	.40064	.40402	68
22	.40402	.40741	.41080	.41421	.41762	.42104	.42447	67
23	.42447	.42791	.43135	.43481	.43827	.44174	.44522	66
24	.44522	.44871	.45221	.45572	.45924	.46277	.46630	65
25	.46630	.46985	.47341	.47697	.48055	.48413	.48773	64
26	.48773	.49133	.49495	.49858	.50221	.50586	.50952	63
27	.50952	.51319	.51687	.52056	.52427	.52798	.53170	62
28	.53170	.53544	.53919	.54295	.54672	.55051	.55430	61
29	.55430	.55811	.56193	.56577	.56961	.57347	.57735	60
30	.57735	.58123	.58513	.59904	.59297	.59690	.60086	59
31	.60086	.60482	.60880	.61280	.61680	.62083	.62486	58
32	.62486	.62892	.63298	.63707	.64116	.64528	.64940	57
33	.64940	.65355	.65771	.66188	.66607	.67028	.67450	56
34	.67450	.67874	.68300	.68728	.69157	.69588	.70020	55
35	.70020	.70455	.70891	.71329	.71769	.72210	.72654	54
36	.72654	.73099	.73546	.73996	.74447	.74900	.75355	53
37	.75355	.75812	.76271	.76732	.77195	.77661	.78128	52
38	.78128	.78598	.79069	.79543	.80019	.80497	.80978	51
39	.80978	.81461	.81946	.82433	.82923	.83415	.83910	50
40	.83910	.84406	.84906	.85408	.85912	.86419	.86928	49
41	.86928	.87440	.87955	.88472	.88992	.89515	.90040	48
42	.90040	.90568	.91099	.91633	.92169	.92709	.93251	47
43	.93251	.93796	.94345	.94896	.95450	.96008	.96568	46
44	.96568	.97132	.97699	.98269	.98843	.99419	1.0000	45
	60′	50′	40′	30	20′	10′	0′	Deg.

NATURAL COTANGENT.

NATURAL TANGENT.

Deg.	0'	10'	20'	30'	40'	50'	60	
45	1.0000	1.0058	1.0117	1.0176	1.0235	1.0295	1.0355	44
46	1.0355	1.0415	1.0476	1.0537	1.0599	1.0661	1.0723	43
47	1.0723	1.0786	1.0849	1.0913	1.0977	1.1041	1.1106	43
48	1.1106	1.1171	1.1236	1.1302	1.1369	1.1436	1.1503	41
49	1.1503	1.1571	1.1639	1.1708	1.1777	1.1847	1.1917	40
50	1.1917	1.1988	1.2059	1.2131	1.2203	1.2275	1.2349	39
51	1.2349	1.2422	1.2496	1.2571	1.2647	1.2723	1.2799	38
52	1.2799	1.2876	1.2954	1.3032	1.3111	1.3190	1.3270	37
53	1.3270	1.3351	1.3432	1.3514	1.3596	1.3680	1.3763	36
54	1.3763	1.3848	1.3933	1.4019	1.4106	1.4193	1.4281	35
55	1.4281	1.4370	1.4459	1.4550	1.4641	1.4733	1.4825	34
56	1.4825	1.4919	1.5013	1.5108	1.5204	1.5301	1.5398	33
57	1.5398	1.5497	1.5596	1.5696	1.5798	1.5900	1.6003	32
58	1.6003	1.6107	1.6212	1.6318	1.6425	1.6533	1.6642	31
59	1.6642	1.6753	1.6864	1.6976	1.7090	1.7204	1.7320	30
60	1.7320	1.7437	1.7555	1.7674	1.7795	1.7917	1.8040	29
61	1.8040	1.8164	1.8290	1.8417	1.8546	1.8676	1.8807	28
62	1.8807	1.8940	1.9074	1.9200	1.9347	1.9485	1.9626	27
63	1.9626	1.9768	1.9911	2.0056	2.0203	2.0352	2.0503	26
64	2.0503	2.0655	2.0809	2.0965	2.1123	2.1283	2.1445	25
65	2.1445	2.1609	2.1774	2.1943	2.2113	2.2285	2.2460	24
66	2.2460	2.2637	2.2816	2.2998	2.3182	2.3369	2.3558	23
67	2.3558	2.3750	2.3944	2.4142	2.4342	2.4545	2.4750	22
68	2.4750	2.4959	2.5171	2.5386	2.5604	2.5826	2.6050	21
69	2.6050	2.6279	2.6510	2.6746	2.6985	2.7228	2.7474	20
70	2.7474	2.7725	2.7980	2.8239	2.8502	2.8770	2.9042	19
71	2.9042	2.9318	2.9600	2.9886	3.0178	3.0474	3.0776	18
72	3.0776	3.1084	3.1397	3.1715	3.2040	3.2371	3.2708	17
73	3.2708	3.3052	3.3402	3.3759	3.4123	3.4495	3.4874	16
74	3.4874	3.5260	3.5655	3.6058	3.6470	3.6890	3.7320	15
75	3.7320	3.7759	3.8208	3.8667	3.9136	3.9616	4.0107	14
76	4.0107	4.0610	4.1125	4.1653	4.2193	4.2747	4.3314	13
77	4.3314	4.3896	4.4494	4.5107	4.5736	4.6382	4.7046	12
78	4.7046	4.7728	4.8430	4.9151	4.9894	5.0658	5.1445	11
79	5.1445	5.2256	5.3092	5.3955	5.4845	5.5763	5.6712	10
80	5.6712	5.7693	5.8708	5.9757	6.0844	6.1970	6.3137	9
81	6.3137	6.4348	6.5605	6.6911	6.8269	6.9682	7.1153	8
82	7.1153	7.2687	7.4287	7.5957	7.7703	7.9530	8.1443	7
83	8.1443	8.3449	8.5555	8.7768	9.0098	9.2553	9.5143	6
84	9.5143	9.7881	10.078	10.385	10.711	11.059	11.430	5
85	11.430	11.826	12.250	12.706	13.196	13.726	14.300	4
86	14.300	14.924	15.604	16.349	17.169	18.075	19.081	3
87	19.081	20.205	21.470	22.904	24.541	26.431	28.636	2
88	28.636	31.241	34.367	38.188	42.964	49.103	57.290	1
89	57.290	68.750	85.939	114.58	171.88	343.77	∞	0
	60'	50'	40'	30'	20'	10'	0'	Deg.

NATURAL COTANGENT.

Special care should be taken in cutting spirals that the work does not slip, and when a cut is made, it is well to drop the work away from the mill while coming back for another cut, or the mill may be stopped and turned to such a position that the teeth will not touch the work while the spiral bed is brought back preparatory to another cut.

Among the spirals most commonly cut on Milling Machines are worms, spiral gears, spiral mills, counterbores and twist drills.

Worms.   In making such cuts as are alike on both sides, for instance, the threads of worms or the teeth of spiral

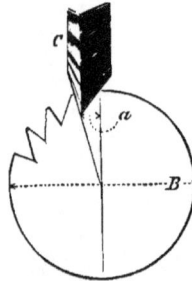

Setting for Spiral Mill Cutting.

gears, care must be taken to set the work centrally perpendicular with the centre line of the cutter before swinging the spiral bed to the angle of the spiral.

Milling Cutters, etc.   Cuts that have one face radial are best made with an angular cutter, as shown on page 127, for cutters of this form readily clear the radial face of the cut, and so keep sharp longer and produce a smoother surface than when the radial face is cut in a vertical plane with a side milling cutter, as shown on page 126, where the teeth can have no side clearance from the work.   The setting for these cuts must also be made before swinging the spiral bed to the angle of the spiral.

By setting the cutter, as shown in above cut, so that the distance a is one-tenth the diameter B, the face cut by the

12 degree side of the cutter will be nearly radial for teeth of milling cutters of the usual proportions.

The operation of forming a twist drill is shown in Fig. 77. The drill is held in a collet or chuck, and, if very long, is allowed to pass through the spindle of the spiral head. The cutter is placed on the arbor so that it will directly over the centre of the drill, and the bed is set at the angle of spiral, as given in the following table :

Twist Drills.

TABLE OF CUTTERS, PITCHES, GEARS AND ANGLES FOR TWIST DRILLS.

Diameter of Drill.	Thickness of Cutter.	Pitch in Inches.	Gear on Worm.	First Gear on Stud.	Second Gear on Stud.	Gear on Screw.	Angle of Spiral.
1/16	.06	.67	24	86	24	100	16° 20'
1/8	.08	1.12	24	86	40	100	19° 20'
3/16	.11	1.67	24	64	32	72	19° 25'
1/4	.15	1.94	32	64	28	72	21°
5/16	.19	2.92	24	64	56	72	20°
3/8	.23	3.24	40	48	28	72	21°
7/16	.27	3.89	56	48	24	72	20° 10'
1/2	.31	4.17	40	72	48	64	20° 30'
9/16	.35	4.86	40	64	56	72	20°
5/8	.39	5.33	48	40	32	72	20° 12'
11/16	.44	6.12	56	40	28	64	19° 30'
3/4	.50	6.48	56	48	40	72	20°
13/16	.56	7.29	56	48	40	64	19° 20'
7/8	.62	7.62	64	48	32	56	19° 50'
15/16	.70	8.33	48	32	40	72	19° 30'
1	.77	8.95	86	48	28	56	19° 20'
1 1/8	.85	9.33	56	40	48	72	20° 40'

The depth of groove in a twist drill diminishes as it approaches the shank, in order to obtain increased strength at the place where the drill is otherwise generally broken. The variation in depth is conditional ; depending mainly on the strength it is desirable to obtain, or the usage the drill is subject to, as in different classes of work. To secure variation in the depth of the groove, the spiral head spindle is elevated slightly; depending, in this case on the length of flute, for which, when 2 inches or less in length, the angle may be ½ degree; 2 to 5 inches, ¾ degree; 5 inches and over, 1 degree. This is generally satisfactory in this respect in our own work, as the drills are seldom very long.

When large drills are held by the centres, the head should be depressed in order to diminish the depth of groove.

The outer end of the drill is supported by the centre rest, as shown in Fig. 79, and when quite small should be pressed down firmly, as illustrated, until the cutter has passed over the end.

The elevating screw of this rest is hollow, and contains a small centre piece with a V groove cut therein to aid in holding the work central. This piece may be made otherwise, to adapt it to special work.

Another, and very important operation on the twist drill, is that of " backing off " the rear of the lip, so as to give it the necessary clearance, to prevent excessive frictional resistance. In the illustration, Fig. 80, the bed is turned about ½ degree as for cutting a right hand spiral, but as the angle depends on several conditions, it will be necessary to determine what the effect will be under different circumstances. A slight study of the figure will be sufficient for this, by assuming the effect of different angles, mills and the pitches of spirals. The object of placing the bed at an angle is to cause the mill E to cut into the lip at e' and have it just touch the surface at e'. The line r being parallel with the face of the mill, the angular deviation of the bed is clearly shown at a, in comparison with the side or the drill.

From a little consideration it will be seen that while the drill has a positive traversing and rotative movement, the edge of the mill at e' must always touch the lip a given distance from the front edge; this being the vanishing point, if such we may call it. The other surface forming the real diameter of the drill is beyond reach of the cutter, and is so left to guide and steady it while in use. The point *c*, shown in the enlarged section, Fig. 80, shows where the cutting commences, and its increase until it reaches a maximum depth at *c*, where it may be increased or diminished, according to the angle employed in the operation, the line of cutter action being represented by *i i*. Before backing off, the surface of the smaller drills in particular, should be oxidized by heating until it assumes some distinct color. The object of this is to clearly show the action of the mill on the lip of the drill, for, when satisfactory, a uniform streak of oxidized surface, from the front edge of the lip back, is left untouched by the mill as represented in the cut at *e*.

We find it a great advantage to grind drills after they are hardened, as they can then be made to run true with the shank. If tapered back about .003″ in 6″ it will be found that this clearance will cause them to run better. To grind the drill it is necessary to make it with a 60° point, as shown in Fig. 80, so that it will run in a countersunk centre. After grinding, this point can be ground off when the drill is sharpened.

It is sometimes preferred to use left handed cutters so that cut will begin at the shank end. By starting the cut in at this end the tendency to lift the drill blank from the rest is lessened.

When giving directions for cutting spirals in any of the foregoing pages, right hand spirals are at all times referred to. For the production of left hand spirals, the only changes necessary are the swinging of the spiral bed to the opposite side of the centre line, the introduction of an intermediate gear upon the stud, Fig. 5, to engage with either pair of change gears for changing the direction of rotation of the spiral head spindle.

Cutting Left Hand Spirals.

FIG. 79.

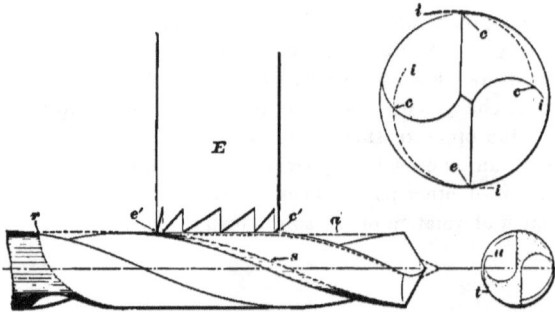

FIG. 80.

Cutters for Making Straight Lipped Twist Drills.

Number of Cutter.	Diameter of Drill.	Diameter of Cutter.	Hole in Cutter.
1	1-16"	1 3-4"	7-8"
2	1-8	"	"
3	3-16	"	"
4	1-4	"	"
5	5-16	2	"
6	3-8	"	"
7	7-16	"	"
8	1-2	"	"
9	9-16	2 1-8	"
10	5-8	"	"
11	11-16	"	"
12	3-4	2 1-4	"
13	13-16	"	"
14	7-8	2 1-2	"
15	15-16	"	"
16	1	2 3-4	"
17	1 1-8	"	"
18	1 1-4	3	"
19	1 1-2	3 1-2	1
20	1 3-4	"	"
21	2	3 3-4	"

List of Cutters for Making Twist Drills.

Number of Cutter.	Diameter of Drill.	Diameter of circle made by Cutter.	Hole in Cutter.	Diameter of Cutter.
1	1-16 in.	.06 in.	7-8 in.	1 3-4 in.
2	1-8 "	.08 "	"	"
3	3-16 "	.11 "	"	"
4	1-4 "	.15 "	"	"
5	5-16 "	.19 "	"	2 in.
6	3-8 "	.23 "	"	"
7	7-16 "	.27 "	"	"
8	1-2 "	.31 "	"	"
9	9-16 "	.35 "	"	2 1-8 in.
10	5-8 "	.39 "	"	"
11	11-16 "	.44 "	"	"
12	3-4 "	.50 "	"	2 1-4 in.
13	13-16 "	.56 "	"	"
14	7-8 "	.62 "	"	2 1-2 in.
15	15-16 "	.70 "	"	"
16	1 "	.77 "	"	2 3-4 in.
17	1 1-8 "	.85 "	"	"

By the use of our special cutters as per list annexed,
Straight Lipped Twist Drills can be made by following the
same general instructions as those just given.

When spirals cannot be conveniently cut with side or angular milling cutters, as previously described, it is sometimes convenient to use end mills, as for example, when the diameter of the piece is very large, or the spiral is of such a lead that the spiral bed cannot be set at the requisite angle, the work is so held that its centre and that of the mill will be in the same plane and the bed is parallel with the ways of the machine. Fig. 81 is a suggestive specimen of work done in this way.

Cutting Spirals with an End Mill.

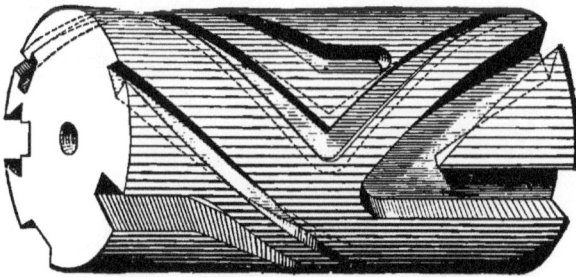

FIG. 81.

INDEX.

Examples of Operations on Milling Machines: